Life in the Hothouse

LIFE IN THE HOTHOUSE

How a Living Planet Survives Climate Change

Melanie Lenart

The University of Arizona Press
Tucson

The University of Arizona Press

www.uapress.arizona.edu

Library of Congress Cataloging-in-Publication Data

Lenart, Melanie, 1961-
Life in the hothouse : how a living planet survives climate change / Melanie Lenart.
p. cm.
Includes bibliographical references and index.
ISBN 978-0-8165-2723-6 (pbk.)
1. Gaia hypothesis. 2. Global temperature changes. 3. Earth. I. Title.

QH331.L5276 2010
550–dc22
2009046053

Manufactured in the United States of America on acid-free,
archival-quality paper containing 100% postconsumer
waste and processed chlorine free.

15 14 13 12 11 10 6 5 4 3 2 1

To Robert Segal
and Mary Roberts,
whose love and support
made this work possible

Contents

Life in the Hothouse

Introduction

The Sweat of the Earth

Sitting at a sidewalk café in Yellow Springs, Ohio, I was surprised to feel a drop of perspiration sliding down my temple. Moving that pen around the page hardly qualified as sweat-producing activity, although I suppose drinking hot coffee on a muggy July day did. Over the next few minutes, I felt tension building—in the air and in me. As the humidity grew more stifling, my body responded by producing bolts of impatience along with beads of sweat. Why couldn't that person keep her dog quiet? And what was wrong with this stupid pen? Suddenly, raindrops began to materialize in front of my eyes, as if arriving from another dimension. I had never before seen air so sodden that raindrops took shape at eye level. I laughed in relief as the oppression of one hundred percent humidity lifted. Every ounce of liquid rain meant that much less water weighing down the air as vapor. As the spattering drops multiplied into a downpour, I grabbed my notebook and retreated inside.

That experience comes to mind now as an apt comparison between two different bodies' natural defenses against overheating—our own and our planet's. As temperatures rise with global warming, we'll sweat more. Similarly, Earth will produce more rain. Both responses have cooling effects. The Earth has other ways of regulating temperature, too. Excess heat spirals into stronger hurricanes and bigger floods. Forests invade grasslands and tundra. Melting glaciers spur the sea to rise up and claim more ground for wetlands. In fact, all these things are happening now. Much as humans produce sweat to cool off on a hot day, Earth produces hurricanes and floods, wetlands and forests to cool off during a hot century. And this promises to be an unusually hot century.

Because of all the greenhouse gases we're producing around the world, the question is not whether the planet's temperature will increase overall—it's how high the mercury will rise, and how quickly. By the end of this century, we face a global temperature change akin to the difference between an ice age and an interglacial warm period. This time around, though, we're already starting from an interglacial warm period. Also, this time around, our own actions will affect how much the Earth warms. At this point, we're not necessarily holding a one-way ticket to hellish heat. Steps we take could slow down the temperature trajectory. They can also moderate local impacts of global warming. Our actions do matter—especially if we follow the planet's lead. The Earth's regulatory system contains some features that can dampen the temperature rise and make it less dire for life, both locally and globally.

The concept that the planet has some means of controlling its temperature comes from Gaia theory, proposed by James Lovelock in the mid-1960s with some later refinement by collaborator Lynn Margulis. With help from the catchy name that refers to the Greeks' Earth goddess, "Gaia" has caught on as an idea that the planet is a living system. For this book, the most relevant aspect of the many-faceted theory involves Earth's ability to keep the global climate within a livable range. In other words, the biosphere—the interconnected web of life inhabiting Earth's land, air, and sea—doesn't just sit around passively, making do with the existing environment. It takes an active role in moderating the environment, including the climate, in ways that promote life's continued survival. Rainfall and other physical mechanisms also contribute to the balancing act. For the past roughly half a billion years, since life branched out to include stemmed plants and insects, Earth's temperature has remained within a livable range. It has shifted between hot and cold without reaching extremes that could turn the world into a snowball or a steambath. Life has clearly played a role in some of these temperature adjustments.

Keep this in mind, though: Gaia keeps conditions livable—not necessarily comfortable. Clearly Earth lacks the precise type of climate-control system found in some modern buildings, with ambient temperature continually adjusted to around 70 degrees Fahrenheit. Temperature can lurch from one extreme to the other, in space and in time. Gaia's climate-control system allows the tropics to persist in never-ending summer and the poles to face relentless winter while the rest of the globe marches through changing seasons. Gaia tends to move from one extreme to the other in time as well as space, as can be seen when all the temperature variations around Earth are lumped together into global averages. Earth's temperature timeline for the past 800,000 years suggests that the planet sometimes crosses a climate threshold—a tipping point—particularly when it comes to the shift from cold to warm periods. Past climate shifts sometimes occurred in a matter of decades, evidence indicates. To be honest, we scientists don't even know exactly where the tipping points for climate shifts are. They're like trapdoors or ejector seats ready to jettison the climate into a different state—and we might not even know we've reached one until after we've passed it.

The idea that the planet has tipping points for climate fits better into Gaia theory than the more standard concept that our world is a mere cog in a mechanical universe. In an eloquent essay on the impact of metaphor in science, philosopher David Abrams suggests a Gaian metaphor should replace Descartes's outdated theory comparing the world to a machine. Comparing the planet to a machine has no more inherent truth in it than viewing it as a living thing, he argued, noting that in many ways the living metaphor better reflects reality. "Gaia, as a self-organizing entity, is no more and no less predictable than a living organism, and we might as well simply acknowledge the fact, and cease pretending it is anything like a machine we could build. The Gaia hypothesis suggests that the world we inhabit is rather more like a living physiology than it is like a watch, or a spaceship, or even a computer," Abrams wrote. While a mechanical model implies that understanding the planet comes down to knowing the right equations, Gaia theory purports that the planet is only as predictable as life itself. And yet we do have some understanding about what conditions allow life to thrive. If we acknowledge the planet as a living system, we can consider global warming in the context of this understanding.

Replacing the imagery of a mechanical sphere with a Gaian living system might be on par with the conceptual leap from basic Newtonian physics to Einstein's theory of relativity and quantum physics. In addition to being revolutionary, both revisions acknowledge that the systems are nonlinear—in other words, they recognize that change doesn't always fall on a predictable line. Sometimes an abrupt change occurs. With the theory of relativity, the rules change at high speeds. Time slows down around the speed of light. In quantum physics, similarly, a process following a nice, predictable line can suddenly shoot off the scale. Quantum physics can help explain why sweating cools us off. As liquid water on our skin collects energy from the sun, it can cross a threshold that shifts it into a completely different phase. One instant it's a gradually (and predictably) warming nano-drop of water. Then, that last quantity of energy arrives, the final photon that propels it into a new frontier.

Suddenly, liquid transforms into gas. Energized, it shoots off into the atmosphere. The drop of sweat has evaporated, along with some of the heat it snagged to fuel its transformation.

Evidence is building that the planet's climate system may have similar thresholds, tipping points that shift it into another state—one that is different yet stable in its own right. Like water, the planet can stabilize in an icy state. Irrefutable evidence for recurring ice ages verifies that. With a thicker blanket of greenhouse gases, including water vapor, the planet can stabilize in a state lacking permanent ice caps. That's what happened during earlier hothouses. Or it can stabilize in an intermediate warm state, as it has for the past eight to ten millennia. Until now. Now climate is changing again.

For the past few decades, there has been a definite and fairly linear temperature rise. It's erratic, sure, but annual ups and downs skitter around a gradually rising line known to scientists as a trend line. Earth warmed by about 1 degree Fahrenheit over the last century, mostly since the mid-1970s. It warmed another 0.4 degree Fahrenheit since just before the turn of this century, as the Nobel-Prize-winning Intergovernmental Panel on Climate Change (IPCC) noted in its 2007 report. IPCC scientists placed much of the blame for the rising mercury on the greenhouse gases we're releasing to power our cars, homes, and factories. The IPCC projects that the globe's average temperature will rise by another 3 to 7 degrees Fahrenheit this century unless we drastically reduce our dependency on coal, oil, and gas. Even if we do make all the necessary changes tomorrow, Earth's temperature is likely to rise by at least another couple of degrees, thanks to the heat stored in the oceans and other global changes already under way.

A temperature rise of 1 degree Fahrenheit or so in a century may not sound like much. During the last century, Earth's average temperature was about 57 degrees Fahrenheit, similar to the average annual temperature of San Francisco. Now it's a little more than 58 degrees, about the average for Greensboro, North Carolina. Yet the 1-degree jump speaks to some big changes in specific areas, in particular years, and in certain seasons.

The U.S. West has been warming faster than the continent as a whole. In Arizona, where I live, average annual temperature has climbed by close to 1 degree a decade, on average, since the mid-1970s. The temperature increase and drought in these semi-arid lands have spurred catastrophic wildfires and large-scale beetle infestations in forests. The number of wildfires larger than one thousand acres increased sixfold in the years after 1985 compared to the previous fifteen years, as a 2006 paper by Anthony Westerling and colleagues documented. In some cases, forest management comes into play, including previous clear-cut logging of big trees and a century of efforts to suppress forest fires. Even so, the scientists analyzing this trend toward bigger western fires found links to spring and summer temperatures high enough to provoke an early meltdown of snow. The earlier snowmelt typically means forests dry out sooner in the season, making them more susceptible to scorching.

The southeastern United States has been lagging behind the rest of the continent in terms of land warming, but many southerners have experienced firsthand the results of warming oceans in the shape of stronger hurricanes. The number of major Atlantic hurricanes more than doubled in the decade ending in 2004 compared to the previous one. Debate continues about how much of the blame for intensifying hurricanes should go to global warming, but no climatologist doubts that higher ocean temperatures can speed up hurricane winds when conditions are right.

Meanwhile, Alaska takes the cake for increase in degrees in the recent warming, with about a 6-degree rise in winter temperatures from the mid-1970s through the early 2000s. "Our snowfall that sticks is coming a lot later. The ice that's thickening is coming a lot later," Hazel Smith, a silver-haired Inuit woman, explained to me over lunch during a 2006 Tribal Lands and Climate conference in Arizona. Not only is the deep freeze of winter coming later in the fall, it's ending earlier in the spring—with repercussions in town and country. Smith's home in Kiana, a riverside town in northwestern Alaska, used to be on the side of a hill. Now it's balanced on top—and the flowing river continues to tear at the hillside soil, which is no longer hardened by permafrost. Her experience is not uncommon among members of her tribe. Entire towns need to be relocated because of eroding permafrost. Along with shelter, the tribe's major food source also is affected. Thinning ice is making it difficult to hunt caribou. "You can't travel where you need to go to get the animals because the conditions are so different now," she sighed. Even when a hunt is successful, a few unusually warm days at the wrong time can spoil the results, as many Alaskan natives count on nature's freezer for meat storage.

IPCC scientists had accurately projected decades ago that warming would occur fastest in the higher latitudes, such as Alaska and Greenland, especially in winter. But no climate models had projected that ice would melt as quickly as recent reports have documented—in part because they did not account for chunks of ice sheets and glaciers breaking away, which accelerates melting. In the summer of 2005, the Greenland ice sheet diminished by about 75 square miles—releasing enough water to supply two hundred sprawling cities the size of Los Angeles for a year. In 2007, the Arctic's normally frozen Northwest Passage opened up for the first time in its known history, allowing ships to pass without using an icebreaker.

Winter gains are not making up for the ice sheet's summer weight loss. The thinning land ice, moving by drips and chunks and entire shelves, is raising sea levels even now, by increments measured in millimeters. Momentum is building, though. Warming water expands, which will add extra height to the sea change. Adding the increase in average sea temperature expected in coming decades to estimates for glacial melt, the IPCC projects the sea level rise will be measured in feet by century's end.

While coastal cities face risks from higher seas and stronger hurricanes, cities everywhere stand on the front lines of global warming. Concrete and pavement play a role similar to that of greenhouse gases, collecting the warmth of the sun during the day to radiate it as heat at all hours. This creates a so-called urban heat island effect that many city dwellers experience firsthand. Thanks to metropolitan Phoenix, Arizona's Maricopa and Pinal Counties have been heating up about three times faster than the surrounding rural areas over the past century, assessments by Arizona State University (ASU) researcher Anthony Brazel and colleagues show. Most of the difference registers after dark, with parts of metropolitan Phoenix simmering about 20 degrees warmer than nearby rural areas at night. Global warming has contributed about a third of the warming, by ASU estimates, while urbanization accounts for the rest of the sizzling temperature rise. Most big cities similarly act as heat traps. Not surprisingly, then, researchers have found that people living in urban areas run a higher risk of heat-related deaths than do their rural counterparts.

This was borne out in the summer of 2006, when about 250 people, mainly in Los Angles and New York, had died in U.S. heat waves by mid-August. Even three hours of cooling during the night can mean the difference between life and death for some members of the most vulnerable population, such as babies, the homeless, and the frail elderly. The lack of nighttime cooling in urban heat islands, then, contributes to making city life more risky during the heat waves that will become more frequent as the world continues to warm. Already, a brutal August 2003 heat wave killed at least 35,000 Europeans, and perhaps as many as 55,000, including about 15,000 people in France. Unlike most French, many Americans have air conditioning. Still, Americans are not as insulated from global warming as we like to think. Most major U.S. power outages occur during summer, right when all those blasting air conditioners are straining the nation's ability to supply electricity. The August 14, 2003, East Coast blackout is a case in point. It left about 50 million residents from Cleveland to New York, and Detroit to Ottawa, without power for more than a day. While England recorded temperatures of 100 degrees Fahrenheit for the first time in its history on August 10 during the deadly 2003 European heat wave, New York City temperatures barely reached 90 degrees during the U.S. power outage four days later. Ditto Cleveland—a fortunate situation, given that its four main water-pumping stations failed during the blackout. As a result, the East Coast blackout was inconvenient but not deadly—this time. As temperatures continue to rise globally, the odds of heat waves becoming deadly will also rise.

Global warming is not the only problem facing society right now, of course, but it is part of an interconnected set of risks that could strain society's ability to respond to problems. The interwoven nature of the global economy became clear when the U.S. economy tanked in 2008, taking many international markets down with it. It wasn't just the people forced into foreclosure by ballooning interest rates who suffered. The financial challenges extended far beyond the failing banks to the world as a whole. Similarly, in a global economy, people everywhere face economic risk from the impact of global warming. For instance, those in the paths of hurricanes featuring an extra punch from ocean warming won't be the only ones paying a price. In the United States and many other countries, people throughout the nation pay for restoration through their taxes (or national debt). Drivers ante up at the pump when damages include oil rigs, such as those toppled by Hurricane Katrina. These costs will join the growing tab picked up by average people as global warming takes its toll. Regional impacts of warming reverberate throughout the country and the world, manifesting in escalating insurance rates, shortages of specific foods, thinly stretched fire-fighting resources, and a lack of National Guard or Federal Emergency Management Agency (FEMA) personnel to help out in back-to-back disasters.

It's a safe bet that most of us would rather not reach a tipping point that topples our financial markets—or the stability of the global climate that supports them. We'd rather not see history repeat itself when talking about the Great Depression. We'd rather not revisit the warm periods that occurred between ice ages, when seas reached high enough to put Miami and Manhattan underwater along with New Orleans. Even 1 to 2 feet of sea rise, the amount the 2007 IPCC report considered likely by the end of the century, would drown neighborhoods and heighten storm surges. What's more, melting ice can set off a chain of events that leads to further warming—less ice cover means less sunlight reflected and therefore more absorbed by the planet. Again, we don't really know at what point that chain of events becomes virtually irreversible, but some scientists are suggesting it may be sooner than we think.

Controversy continues to rage over how our recent warming compares to temperatures of the past century or millennium. In November 2009, the unauthorized release of hacked email exchanges among a few prominent climatologists led some skeptics to renew their challenges of data indicating that the global temperatures of the 1990s stand out in long-term records. In December 2009 the World Meteorological Organization raised the bar by reporting that the 2000s surpassed the 1990s as the hottest decade, in this case since the instrumental record began in 1850. In a way, though, the wrangling over which year or decade of the past century or thousand years was hottest overlooks the real issue: Why was it so hot, then or now? When we shift our sights past the superlatives and go even deeper into the past, we find a consistent—if not exactly predictable—link between high greenhouse gases levels and high temperatures.

Many uncertainties exist when trying to forecast future climate, whether next season's rains or the coming century's temperature rise. We can acknowledge these uncertainties while embracing the advances in understanding gained over years of dedicated research by hundreds of scientists. Here, I make an effort to describe the uncertainties where they exist in various records, from ice cores to marine deposits to fossil plants, with the hope that readers will then have the information they need to draw their own conclusions, which may be quite different from mine. At the same time, I think it would be a stretch to conclude from the evidence described here that there's no cause for concern.

Life can survive in the face of global climates much colder and much warmer than today's relatively mild climate. In fact, our planet has spent much of its existence warmer than today, even counting the extra heat from the current global warming. Not every species survives changing climate, though. What's more, we need to go back hundreds of thousands and even millions of years to find episodes in the geological record resembling what we might expect in coming decades. This book focuses on the past 100 million years, including time slices during two major hothouses, the Cretaceous and the Eocene. It also explores changes during the Quaternary Period of the volatile last couple of million years or so, when climate regularly cycled between ice ages and warm periods. These lessons from the past have much to teach us about our situation today and in the near future—including how we can work with the planet to limit some of the potentially disastrous impacts of global warming. Anything we can do to defuse the temperature climb or reduce its impacts will help humans and the other species that help shape this spherical living system we call Gaia, whether as stewards of the planet or passengers just trying to survive and thrive.

In his 2006 book, *The Revenge of Gaia*, James Lovelock, who was then eighty-seven, described Gaia as an "old lady," frail and potentially unstable as she nears the end of her life. (About 3.5 billion years down, only a billion to go.) On the other hand, in her 1998 book, *Symbiotic Planet*, the indomitable Lynn Margulis called Gaia "a tough bitch" who has managed to survive for billions of years and faces no risk from mere humans. In my mind, Gaia is something in between—perhaps a middle-aged planet suffering from a greenhouse-gas attack. Life on this planet will certainly survive the vagaries of this human-caused condition—but many species and individuals might not. Unlike Lovelock, I don't really see humans as one of the threatened species. But I do worry about how many individual people will suffer from the changes related to the ongoing warming.

For humans and other life forms to thrive, care must be taken so the planet's overheated condition doesn't worsen and become irreversible. Gaia theory brings home that we are not just living on the rocky, watery surface of a planet. We are living, breathing parts of a living, breathing planet. If that planet catches a cold, or burns with a fever, we also suffer. The sooner the planet returns to a healthy state, the sooner we also can breathe easier. This is especially true if we acknowledge that Gaia cools off not only through rainfall but also through stronger storms, including hurricanes.

1

A Feverish Response

Hurricanes Come with High Temperatures

Huddling under an overturned couch while Hurricane Hugo raged over Puerto Rico, I wondered why on earth hurricanes existed. A native of Chicago, I had never experienced the shrieking winds and incessant rains that come with major hurricanes. Tornado warnings faded away within hours, and I had never landed in the path of these relatively skinny twisters. Hurricanes stretch across hundreds of miles with rains that can last for days. I hoped upending the couch would protect me from a potential ceiling collapse. Dozens of hours, millions of toppled trees, and billions of raindrops later, the storm had passed. The San Juan area fared surprisingly well, although things would have been different if floodgate glitches that occurred had led to the Carraizo Dam's collapse. If the dam had collapsed, tens of thousands of us might have been threatened by floodwaters, later reports revealed. Instead, the tropical rain forest on the island's northeastern end absorbed the brunt of the storm's force. This experience and the heartening revival of the tropical forest over the next few months impressed me. Hurricanes stayed on my radar throughout my subsequent studies of forests, rivers, and global climate change.

It may make sense for a warming world to produce stronger hurricanes as temperatures rise, somewhat as people might sweat on a hot day. Greenhouse-gas emissions by society have been heating things up, so it's logical to expect Gaia to respond with efforts to maintain climatic equilibrium. To reiterate, Gaia theory holds that over the long term, our living Earth—Gaia—has ways of keeping its climate within a range suitable for life. Could hurricanes play a role in tweaking the planetary thermostat? Unfortunately for those of us who end up in their paths, my subsequent research and conversations with experts indicate that this cooling function helps explain their existence. Exploring some of the reasons for this claim can offer an introduction to one way of viewing planetary dynamics through a Gaian filter. This particular Gaian perspective adds an element of reverse engineering—that is, an attempt to work backward from planetary responses to possible values for those responses.

❧

Hurricanes have a regulatory reason to exist when it comes to planetary cooling, for both physical and biological reasons. Let's start with the physical, always a good idea when considering planetary health. Although hurricanes may look like a way for a vengeful planet to let off steam, they actually cool the sweltering tropics by redistributing heat to other regions. This cooling function helps explain their existence. Hurricanes arise in the tropics, generally between 6 degrees and 20 degrees from the equator. (Closer to the equator than 6 degrees, the Earth's rotation doesn't impart enough spin to bring them into circulation.) After they mature, they're more able to withstand the cooler seas and stronger wind shear outside of tropical latitudes. By definition, the tropics receive a direct blast of sunlight at some point in the year. These spot-on rays contain more energy per square foot at Earth's surface than the angled rays reaching areas where the globe curves more blatantly. Consider a flashlight beam: If you aim it directly at a spot on the wall, it shines brightly on a small area. If you angle it to cover more ground, a weaker beam sheds light on a larger area. The tropics receive direct beams, some of them several times stronger per square foot than those striking Canada. This focused energy warms seas, makes clouds, and spawns hurricanes. All of these processes use heat.

The tropics are net exporters of heat, sending it off via winds, ocean currents, rain—and, of course, hurricanes, which employ all three methods of shifting heat elsewhere. It takes a lot of energy to warm water and to turn it into vapor, energy that could otherwise heat the air. Wind and heat often carry this captive energy up and away, until it escapes from its bonds when water vapor condenses into raindrops or snowflakes. Likewise, hurricanes cool the air by converting rising heat into the kinetic motion of swirling winds. Hurricanes also push heat downward and poleward as they surf the waves. Through basic physics and its interactions with its environment, hurricanes serve to cool the tropics. It's only logical, then, to expect a warmer world to feature more intense hurricanes.

Planetary mechanisms seem to support this logic. It's no coincidence that hurricane season typically starts as summer peaks and fades away by the winter solstice. High temperatures at the ocean's surface kick hurricanes into action. Other rules apply, too, especially in the hurricane's formative stage. High-level winds must be calm enough not to shear off the storm's head of steam. Hurricanes also need a lot of humidity. The humidity issue helps explain why dusty air can suppress hurricanes, as it may have done during the unexpectedly mild 2006 Atlantic season. Once an intense hurricane is off and spinning, though, sea-surface temperature seems to drive or at least predict its power surges, as hurricane expert Kerry Emanuel of the Massachusetts Institute of Technology has found. When explaining a hurricane's intensity, Emanuel found no need to invoke wind shear or any of the other factors affecting hurricane formation. Knowing the ocean temperatures around the hurricane eye gave him enough information to mathematically mimic most of the ups and downs of several intense hurricanes he modeled in a 1999 *Nature* paper.

Surface waters must reach about 80 degrees Fahrenheit to get these storms rolling. Waters topping 83 degrees seem to bring on more intense Atlantic hurricane seasons. This general pattern also comes through at the scale of individual storms encountering warm waters. That's not to say that every hurricane strengthens when it hits warm water. Other factors do come into play, most notably wind shear. But there's little chance a hurricane will attain major league status without passing over warm water.

Even hitting a pocket of warm water can amplify hurricane winds, if atmospheric conditions don't interfere. That's what Lynn "Nick" Shay of the University of Miami and others reported in a 2000 *Monthly Weather Review* paper describing their findings on the Gulf of Mexico's 1995 Hurricane Opal. This storm leapt from Category 1 into Category 4 in just fourteen hours while crossing a roaming pocket of deep warm water. In 2005, Shay and some of his colleagues flew right over Hurricane Rita to test the waters in the Gulf of Mexico. Winging into the eye of the storm, they dropped measuring devices into the churning depths. Their sea-tossed thermometers continued reporting temperatures as they sunk, to depths as low as 5,000 feet. With these efforts and others, they documented that 2005 Category 5 Hurricanes Katrina, Rita, and Wilma intensified while passing over warm waters related to the Loop Current—a tongue of deep, warm water lapping into the Gulf from the Caribbean. "All three of those storms interacted with the warm current—the Gulf of Mexico Loop Current—and all three of them exploded," Shay said. "You can actually think of these deep warm pools as fuel injectors."

Swimmers know that surface waters generally run the warmest, whether it's an ocean or a lake or even a pool. The deeper the warmth penetrates, the more potential it has to rev up hurricanes. Although hurricanes don't create whirlpools, their winds thoroughly stir up the ocean beneath them. They also displace water, pushing it ahead as a formidable storm surge. "Even out in front of the storm, the winds are already mixing the oceans," Shay noted. "By the time the eye of the storm is close by, it's really feeling the mixed layer temperature."

A shallow layer of warmth on a deep, cool sea soon turns into a lukewarm mix, with little energy to power hurricanes. This helps explain why eastern Pacific hurricanes often drop down to tropical storm status if they head north. The Pacific region off Baja California actually launches more hurricanes than the coast of Africa, the birthplace of many Atlantic hurricanes, but these eastern Pacific hurricanes get cold feet if they head north. A frigid current from Alaska soon cuts off their charge. (Of course, that current could warm as Alaska's temperatures continue to rise.) In contrast, Atlantic hurricanes can ride the Gulf current to reach as far north as Canada, though rarely with an intact eye and the extra punch that comes with it.

❧

There's a flip side to this churning action of hurricanes that hints at a Gaian role. When hurricane winds mix the ocean, they leave a trail of cooler waters in their wake. That's what atmospheric scientist Morris Bender and oceanographer Isaac Ginis documented in a 2000 paper. Hurricanes tangibly cool the sea surface around them through evaporative cooling and by mixing in cooler waters from depths where the sun doesn't reach. A hurricane's influence on the ocean beneath it easily extends down 300 feet, and up to 1,000 feet in some cases. Meanwhile, the sun's rays typically penetrate about 300 feet at best. (Of course, warm currents like the Gulf Stream can provide another source of heat.) In a computer animation of Katrina's path that Bender and Ginis posted on the Geophysical Fluid Dynamics Laboratory's Web site, the sea in front of Katrina shows up in blazing red, with surface temperatures registering in the mid-80s in degrees Fahrenheit. Behind the storm, temperatures fade to a cool blue, registering in the mid-70s and below. This surface cooling can last a week or more, potentially affecting future storms, Ginis noted. For instance, when Hurricane Danielle in 2004 crossed the same path used by Hurricane Bonnie four days earlier, Danielle's winds dropped from about 80 miles per hour to 65 miles per hour.

Bender, Ginis, and another colleague compared more than a dozen hurricanes and found that their ability to cool the surrounding sea related directly to the time they lingered over an area. Loitering hurricanes cooled the surface by an average of 10 degrees Fahrenheit, whereas fast-moving hurricanes cooled it by about 3 degrees Fahrenheit. Hurricane Georges dawdled in the Gulf of Mexico long enough to cool sea-surface temperatures by up to 6 to 9 degrees Fahrenheit in its wake, making its path to the southern coast (including Louisiana) visible for weeks using temperature-detecting satellites. Thanks to satellites that can sense infrared heat, it's easy to see that hurricanes cool the ocean surface along their paths.

Can hurricanes effectively cool the sea surface from one year to the next? Maybe this would help explain why the 2006 Atlantic hurricane season thankfully fizzled out almost before it started. The 2005 Atlantic season produced twenty-eight named tropical storms, a high in the more than fifty years of reliable records. The 2004 season was no slacker, either, yielding fifteen named storms, including a record four hurricanes striking Florida in one season. In contrast, only five named Atlantic storms formed in the 2006 season. Other factors also intervened, including an El Niño event. In a 2007 *Scientific American* article, climatologist Kevin Trenberth linked the relatively mild hurricane season to stronger wind shear during a 2006 El Niño event as well as cooler sea-surface temperatures that season, which he related to heavy trade winds during the preceding winter. Theoretically, the previous hurricanes also could have contributed to cooler seas in 2006. Perhaps the 2004 and 2005 Atlantic hurricanes helped export enough heat from the tropical Atlantic to prevent another catastrophic hurricane season on their heels.

This idea is speculative, but not without support. A 2007 analysis by Ryan Sriver and Matthew Huber of Purdue University indicates that hurricanes acted to cool the surface of many cyclone-prone seas during the period they considered, between 1982 and 2001. They found a pronounced cooling in the Gulf of Mexico and the Atlantic from the Caribbean north up the U.S. East Coast to Canada. Much of Asia's east coast also registered extensive cooling, as did Southern Hemisphere pockets favored by hurricanes. This cooling occurred even as the world's oceans warmed overall along with the planet as a whole. Estimating the surface cooling to 150 feet deep in hurricane-prone regions, Sriver and Huber figured hurricanes could account for about 15 percent of the heat moving out of these regions in a typical year. Huber, an associate professor at Purdue who often works from a laptop in the local Café Vienna, explained by e-mail that the climate-change community has expressed strong interest in his research, but most oceanographers still resist the concept that short-lived storms could redistribute so much heat. "No one has thrown tomatoes yet, so that's good, but I think the community is still skeptical, which is fine," he added.

Still, scientists have long recognized that some line items are missing in the tropical heat budget. The current numbers do not account for all the heat known to move from the tropics to the subtropics. Meanwhile, Sriver and Huber's research shows that the more intense storms move the most heat. As Huber put it, "Strong storms lead to more ocean mixing than weak storms that last longer." A 2008 paper of theirs with colleague Jesse Nusbaumer, based on higher-resolution information from the tropics for the 1998–2005 period, suggests the amount of surface cooling occurring could be even about a third higher than their earlier estimates. In response to my question about whether the hurricane-cooled surface might help explain the relatively mild 2006 Atlantic hurricane season, Huber responded: "I think that's possible, but by no means proven. It's conceivable that a huge amount of heat was flushed downwards (and cold mixed up), but we may see that heat back again." Emanuel's work, some of which is described in a 2001 paper, suggests that most of the heat is lurking in deeper waters poleward of its tropical source.

Even while cooling the tropical oceans with their winds, hurricanes also cool the tropical air and sea by promoting evaporative cooling and by exporting rain. In fact, rainfall processes associated with hurricanes moved about three times more heat out of the tropics than ocean cooling, estimated Kevin Trenberth and John Fasullo in a 2008 paper. Hurricanes will pull in air moisture from up to 1,000 miles away. All the rainfall that comes down in days-long downpours or drizzles started out as water somewhere on the surface. And the evaporation of water lifts heat from a surface—whether skin, land, or sea. The thunderstorms that shape hurricanes, forming arms of convective clouds that stretch far from the hurricane's eye, arise specifically because of heating at Earth's surface. The heat transforms into another form of energy, in this case the energy that turns earthbound water into airborne gas. (Remember, old energy never dies, it just changes form.) At some point, that gaseous water vapor will reach a point where it can no longer sustain its free-floating lifestyle. Often, that moment occurs after it has departed the warm confines of the tropics. When it heads back down to earth, it releases the energy it collected to fuel its journey. Similarly, hurricanes themselves wind down at some point, especially after departing the energizing tropics. The winds and rain they carried with them share energy and heat with distant lands. But in their tropical stomping grounds, and anywhere they use latent heat from evaporation to fuel their powerful winds, hurricanes are helping to counteract the buildup of heat that comes with global warming.

❧

Hurricanes evolved as a cooling mechanism long before people built cities in their paths. Even so, it's easy to take a hurricane personally in the heat of the action. Raging winds, crashing trees, roaring rivers—even the descriptions of what happens inside a hurricane bring to mind an angry god or goddess. Perhaps it is the god Hurakán, a pre-Columbian deity of the Tainos, the Caribbean islanders who passed down the western word for these winds of destruction. Or maybe Kali, the Hindu goddess of destruction with her many arms. In most of Asia, these swirling storms are called typhoons, which some relate to the Cantonese phrase *tái fung* (great wind). By any name, tropical cyclones are all the same at the core: low-pressure systems surrounded by high-speed winds swirling around an inner circle, with multiple stretches of thunderstorm clouds circulating in the outer reaches. And they're big, typically stretching hundreds of miles across. Tropical cyclones that sustain wind speeds of 74 miles per hour or more qualify for hurricane or typhoon status based on the commonly used Saffir-Simpson hurricane intensity scale. In this book, I'll use "hurricane" to describe all tropical cyclones that cross this wind-speed threshold. It seems fitting to follow the Taino tradition, as I experienced two major hurricanes on the Caribbean island the Tainos called Boricua.

My second encounter came about a week after I returned to this enchanting Caribbean island, now known as Puerto Rico (although to this day, islanders often refer to themselves as Boricuas). Following a two-year stint at graduate school, I had come back in September 1998 to collect information on uprooted trees for my dissertation. A few days after my arrival, word got around that an Atlantic storm was headed in our direction: Hurricane Georges (pronounced "zhorzh"), presumably named in honor of French-speaking Caribbean islanders. Visions of uprooted trees ripe for the measuring made me perhaps one of the only islanders with positive thoughts about the coming event. Somewhere inside of me, a scientist rubbed her hands together in anticipation. I was yanked back to the reality of being human by some news that arrived with my breakfast on the morning before the hurricane's projected landing. Georges's sustained winds had reached 150 miles per hour. Another 6 miles per hour would push it into Category 5 status—the highest ranking on the Saffir-Simpson scale of destructive power. Fear thudded into my stomach, destroying my taste for scrambled eggs. There was no turning back now. It's impossible to evacuate an island of nearly 4 million people, even with several days' notice. So we boarded up windows, stocked up on canned soup, batteries, and beer, and waited. At least this time I would share the experience with friends in the forest instead of alone in an urban rental, where I had spent much of Hurricane Hugo when it roared past San Juan.

While the electricity lasted, we watched the satellite images of this Texas-sized cyclops navigating the Atlantic at 20 miles per hour—the typical slow-motion lurch of hurricanes. Georges stretched several times the size of Puerto Rico's 100-mile length. At times, the eye diameter alone matched the island's 35-mile width. That's bad news. Although people use "eye of the storm" to describe the relatively cloud-free and calm center, the eyewall circling it marks the most ferocious winds of the beast. Luckily, before Georges cast its eye upon Puerto Rico, its sustained wind speeds dropped to about 115 miles per hour, placing it into a more manageable Category 3. My friends and I spent some of the night reinforcing a wall of their wooden home in the mountains outside of Luquillo, near the hurricane's point of entry on the island's eastern end. But we remained safely sheltered throughout the storm.

During the eye's eighteen-hour passage across the length of Puerto Rico, Georges blew off roofs, uprooted trees, launched landslides, and flooded cities. Two days of rainfall totaling 28 inches devastated the mountain town of Jayuya. Four out of five of the wooden structures crumbled on the associated islands of Vieques and Culebra. Some 73,000 homes in Puerto Rico were destroyed or damaged, like a cliffside home we drove by that was filled with mud from a landslide. Although no one in Puerto Rico died directly from the hurricane, several died in the aftermath. In the Luquillo Mountains, we drew our electricity from noisy generators for about a month and washed in the nearby Sabana River while local teenagers treated the powerless wires overhead as jungle-gym equipment. Sudden surges of electricity into these dormant cables caused a few deaths around the island in the weeks after the storm. Still, by the time I left Puerto Rico nearly four months later, life had returned to a semblance of normality. Traffic moved in its usual erratic patterns, splintered forests sported a new set of greenery, and carpenters labored to repair the damaged structures. People went about their business.

In contrast, vast expanses of New Orleans still looked like ghost towns four months after the August 2005 strike from the more intense Hurricane Katrina. A weeklong stay in December of that year to document the impacts and lend a hand in relief efforts put me amid the lingering devastation. Piles of furniture and debris lined the sidewalks in front of homes that had been soaked by floods. Blue tarps flagged damaged roofs throughout the city. Knee-high signs sprouted around traffic stops advertising home-gutting services, lawyers, and toll-free numbers to call for quick home sales. In the Lower 9th Ward, most commercial and residential streets looked deserted, with the occasional family arriving for days or weeks to check damages and consider whether to attempt repairs.

It was easy to see why so many structures remained abandoned despite the typical façade of normalcy on the outside. The exterior mud lines from the waist-high floodwaters had faded away, but the inundation had allowed toxic black mold to insinuate itself into the interior woodwork and walls, along with a host of other lung-damaging substances. Survivors had a name for the trouble with breathing the local indoor air: the Katrina Cough. All these flooded homes required gutting, a dangerous task even with the respirators and space-age suits worn by the brave souls who ventured inside. During my brief stint with the relief group Common Ground, these courageous volunteers included more than a hundred people willing to gut houses by day and sleep fifty to a room by night in the Baptist church the group was in the process of restoring at Pauline Street and Claiborne Avenue.

Down the street from the Common Ground distribution center, I met Kirk Armelin, who agreed to tell me his story. "This is the house I was raised in since '66 or '67," Armelin said, gesturing back to a pink bungalow that maintained its exterior charm. "We get hurricanes like they get blizzards up north. I've been through Betsy, Camille—none of them never did nothing. The furthest the water's ever been is the third step right here," he said, pointing with a broom to a step about a foot off the ground. "But this one took a toll. This one took everything." Like many in the neighborhood, his house filled halfway up with a toxic stew of muddy water during Katrina. Versions of Armelin's story could be repeated 150,000 times over in the city known as the Big Easy. That was the rough tally for the number of Katrina-damaged homes in the New Orleans area alone.

Right next to the 9th Ward, the French Quarter's higher ground and stucco construction helped its ancient buildings endure several centuries' worth of major hurricanes, including Katrina. Still, problems arose in this eighteenth-century neighborhood from the wind, lack of electrical power, and lawlessness that prevailed for days and weeks following the hurricane. The power outages did their own damage, as I learned from Market Café owner John Tsatsoulis. Four months after Katrina hit, Tsatsoulis was still struggling to get his French Quarter business reopened. The day I met him, he was having new commercial refrigerators and freezers installed to replace the ones destroyed when rotting food formed acid leachate during the power outage. While we sat at the empty circular bar with its expansive view of Decatur Street, Tsatsoulis sadly showed me pictures of what used to be his home in Lakeview, the neighborhood bordering Lake Pontchartrain. A storm surge on top of rains had overpowered the 17th Street Canal and the resulting breach inundated many homes in this formerly ritzy neighborhood. In the Tsatsoulis residence, water rose past the gutters, weakening the structure enough to pull down the ceiling. After we looked at the images of his former home as a pile of debris, he shook my hand warmly and said he appreciated research on how these problems relate to global warming.

❦

Atlantic hurricane damage and high temperatures went hand in hand in both 1998 and 2005, the world's two hottest years in the instrumental records (at least as of publication). This does not prove anything. It could well be coincidence. Still, rising sea-surface temperatures relate to both. Warm oceans helped push 1998 into the record books as the world's hottest year in at least a century—and possibly in a millennium, when instrumental records are compared to records developed from tree rings and other natural archives. The year 1998 also made the record books for highest insured losses from weather-related events around the world, estimated at $92 billion. About a month after Hurricane Georges menaced Puerto Rico and the Gulf Coast, the even more powerful Hurricane Mitch crossed the Atlantic to reach Central America. High sea-surface temperatures accelerated wind speeds of both Georges and Mitch. Damage tallies cannot begin to account for the cost of the storms. Mitch killed about 11,000 people when it plowed through Honduras, Nicaragua, El Salvador, Guatemala, and Belize. After passing over a section of warm ocean current, Hurricane Mitch's sustained winds reached an estimated 180 miles per hour out at sea. It dropped to about 150 miles per hour—just below the Category 5 threshold—right before its eye breached land in Honduras. Of course, half the storm preceded the eye, and Emanuel's work has shown that conditions right around the eyewall matter most to hurricane dynamics.

Less than a decade later, 2005 challenged 1998 as the warmest year in the instrumental record. Different analytical approaches yield different results for which year was hotter, but they were both scorchers. And 2005 brought the Atlantic three Category 5 storms, featuring Hurricanes Rita and Wilma as well as Katrina. The 2005 hurricanes went on to break U.S. records for insured costs as well, according to the National Hurricane Center's report on Katrina. The cost in lives was high as well. More than 1,300 people died throughout the South, with at least 1,000 from Louisiana, more than 200 in Mississippi, and the rest in Florida, Georgia, and Alabama. A shamefully deficient emergency response made things worse. But there's no question that the risks and challenges become much greater as hurricanes gain power.

The difference in destructive power among hurricanes is tough to capture with the Saffir-Simpson five-category scale. In reality, the categories make near-quantum leaps in ferocity. On my December 2005 flight from Colorado to New Orleans via Memphis, I sat next to a fourteen-year-old girl, Mary Katherine, who recalled playing in the waves among surfers when the Category 2 Hurricane Dennis passed near Jamaica while she and her family were vacationing there earlier that year. You might think Dennis's 50-mile-per-hour winds would carry about one-third of the destructive power of Mitch's 150-mile-per-hour winds. But no. With the exponential difference factored in, Mitch contained roughly *twenty-seven times* the destructive power of Dennis, with tragic results. Hurricane Katrina's destructive power fell closer to Mitch's than Dennis's on the spectrum. The 134-mile-per-hour gusts measured in Poplarville, Mississippi, during Katrina delivered nearly twenty times the destructive force of Dennis's winds. Katrina also heaved in a wall of water to flatten many homes in Gulfport, Mississippi, where Mary Katherine lived. She and her family were headed home to rebuild their lives after visiting relatives in Colorado. "At least we got away from the destruction for Christmas," the lanky brunette said with a philosophical shrug.

When Katrina's eye breached land at Buras, Lousiana, its sustained winds of about 126 miles per hour fell just under the 131-mile-per-hour speeds required to qualify for a Category 4 storm. But because a hurricane's destructive power increases exponentially with wind speeds, Katrina contained about twice the destructive power of Hugo's 92-mile-per-hour winds in San Juan, even though both qualified for Category 3 status. (Hugo qualified because its winds were stronger around the eye, which crossed over the island's northeastern corner.) While Hugo's 1989 efforts to burst through the island's Carraizo Dam floodgates failed—just barely—Katrina's extra punch gave it a final surge of power that tipped the scales: Its floodwaters broke through several crucial levees just when residents were beginning to hope that the worst of the storm had passed. This led to destruction so widespread it remains to be seen whether parts of this beloved Louisiana city will ever recover, especially in "the bowl," as locals call the dip in the land so clearly delineated in the flooding that followed in Katrina's wake. This is the area where so many of the city's people lost their lives—linked in part to an emergency response that was widely acknowledged to have been a disaster in its own right.

Many Katrina survivors said their insurance companies were refusing to pay for flood damage even though they had hurricane coverage. News reports concurred with their impressions, suggesting insurance would cover only half or less of the estimated $100 billion in damages. Yet, from a scientific perspective, floods fall squarely within the realm of hurricane destruction. Hurricane category generally predicts the heights of accompanying storm surges. For instance, Katrina pushed a wall of seawater onshore that reached 27 feet high at Hancock, Mississippi, and stretched inland for 6 miles—as befitting the Category 5 status it held a mere twelve hours before its eye reached land. Most of the people killed by Katrina died in floods. Similarly, most of the deaths from Hurricane Mitch's 1998 rampage through Central America resulted from floods and related landslides. The mountainous terrain that helped break up Mitch's eye and slow wind speeds also prodded the storm into unleashing torrents of rainfall. An estimated 50 to 75 inches fell on some towns, washing out hillsides and damaging millions of homes. Flash floods and mudslides also killed most of the 600 people who died in Hurricane Georges, mainly in the Dominican Republic and Haiti. In fact, floods, rather than winds, have caused most hurricane deaths documented over the past few decades.

Heavier rains come with stronger hurricanes, as Arizona State University geographers Randall Cerveny and Lynn Newman found when they compared nearly 900 tropical cyclones using satellite imagery. Every 22-mile-per-hour increase in sustained wind speeds brought an extra inch of daily rain on average, based on their analysis. Given the documented tendency for hurricane rains to increase with winds, climate modelers Thomas Knutson and Robert Tuleya estimated that hurricane rainfall would increase by about 18 percent by mid-century given the projected warming of tropical oceans. Rainfall tends to come in more intense storms in a warmer climate (as chapter 4 will describe). That doesn't mean every extra hot year will bring more hurricanes and floods. It does mean the odds of stronger hurricanes accompanied by bigger floods are going up along with sea-surface temperature. As Robert Corell, chairman of the Arctic Impact Assessment, put it during a keynote talk at a 2006 Tribal Lands and Climate conference in Arizona, "Katrina is consistent with the fact that we will have more Category 4 and 5 hurricanes. As the planet warms—it doesn't take much warming—the volatility of the system increases."

❧

Warmer seas already are wielding stronger hurricanes around the world, as Peter Webster of the Georgia Institute of Technology and some colleagues reported in 2005. The team of four scientists estimated changes in hurricane strength using records from satellite imagery going back to the mid-1970s. These records cover the oceans and other locations lacking wind-measuring devices. The research team found an astounding jump in "intense" hurricanes—those with sustained wind speeds over 130 miles per hour—when comparing the first half of the satellite record with the second half. A full 269 intense hurricanes thrived during the fifteen-year period that ended in 2004, a 57 percent increase over the previous fifteen-year period. The difference averages out globally to an extra six intense hurricanes a year. Meanwhile, minor hurricanes had declined in number. The typical 80 to 90 hurricanes were still forming each year as before, but during the warmer period, more of them reached "intense" status. The satellite data remain controversial to some researchers, as do other less comprehensive records. Based on existing data, which are imperfect but available, a 2006 study by Ryan Sriver and Matthew Huber reached similar conclusions for individual ocean basins included in the analysis.

Also in 2005, mere weeks before Katrina struck, Kerry Emanuel had published similar results for the Atlantic and North Pacific—something he said surprised even him. Using observational records from 1949, he found that Atlantic hurricanes had nearly doubled in power since about the mid-1970s. As forecasters often do, Emanuel combined storm length and intensity into a value reflecting total destructive power of regional tropical storms (including hurricanes) for each season. Power rose and fell with average sea-surface temperature, both seasonally and across the decades. Since his findings were published in 2005, the distinguished Massachusetts Institute of Technology professor has been sounding the alarm about how global warming is strengthening hurricanes. Atlantic hurricane activity seems particularly in tune with sea-surface temperatures, more so than hurricanes in some other ocean basins, as indicated in his 2008 journal article written with colleagues. In 2006, Emanuel and lead author Michael Mann published an *Eos* article showing that Atlantic ocean temperatures tend to move up and down in sync with air temperatures in the Northern Hemisphere, at least when averaged by a decade at a time. The results imply that as long as greenhouse gases continue to heat up the Northern Hemisphere, the Atlantic will continue to spawn intense hurricanes.

Critics of the theory that greenhouse gases are involved in warming the hurricane-prone Atlantic argue that natural variability of the ocean currents might explain the jump in hurricane power, and they question whether the observed increase is real or merely a result of improved detection in recent decades. Those favoring natural variability and improved detection include respected National Hurricane Center researchers William Gray and Christopher Landsea. In 2009, Landsea was the lead author of a *Journal of Climate* paper touted as showing that the observed increase in the number of tropical storms and hurricanes in the past century related to improved detection. The authors concluded that a perceived increase in the number of "short-lived storms"—that is, those lasting two days or less—related to improvements in scientists' ability to detect these storms with modern technology. A co-author on that paper, however, pointed out in an August 11, 2009, press release issued by the National Oceanic and Atmospheric Administration that the study did not consider whether Atlantic hurricanes had strengthened in recent times. As co-author Thomas Knutson noted, the study "does not address how the strength and number of the strongest hurricanes have changed or may change due to global warming."

As far as variability goes, all climate researchers recognize the challenges of distinguishing between an upward trend and a natural fluctuation from a thirty-year or even a fifty-year record. Like any climate pattern, hurricane intensity also waxes and wanes as part of natural climate variability. Collective hurricane power varies by year and even in fluctuations stretching across decades. In fact, Gray and Landsea were among those expecting an eventual rise in Atlantic hurricane power in the 1990s based on decades-long fluctuations of regional sea-surface temperatures and other factors. When it comes to natural variability, though, Atlantic hurricane activity tends to increase while activity decreases in other regions, most notably the eastern North Pacific. In contrast, the satellite-based analysis by Webster and his colleagues showed all six ocean basins registering an upswing in the number of intense hurricanes since 1990. What's more, the timing of the upswing echoed a documented rise in sea-surface temperatures of the same ocean basins—a rise linked to global warming from burning gas, coal, oil, and forests.

Gray maintains that a recent increase in Atlantic hurricane destructive power relates to natural variability rather than global warming. Perhaps this isn't surprising—he doesn't believe that society's emissions of greenhouse gases are warming the planet, as Chris Mooney describes in detail in his 2007 book *Storm World*. Still, decades ago, Gray identified sea-surface temperatures over 80 degrees Fahrenheit as one of the key ingredients in nature's recipe for hurricane formation. Sea-surface temperatures fluctuate based on factors besides global warming, too, of course. They move up and down with the comings and goings of ocean currents, from natural variability, and even from exposure to major hurricanes themselves, as described above.

Another major factor affecting a storm's ability to attain and retain major hurricane status involves vertical wind shear—the ability of high-level winds to bend hurricanes out of shape. As Landsea has noted, some computer models project that increased wind shear might largely offset the effect of warming seas. For instance, Gabriel Vecchi and Brian Soden compared eighteen global climate models and found that many projected an overall increase in wind shear over the tropical Atlantic and eastern Pacific. On the other hand, Landsea and Gray, writing with lead author Stanley Goldenberg and another colleague, reported a connection between above-average Atlantic sea-surface temperatures and below-average wind shear in a 2001 paper. A 2008 research effort led by Chunzai Wang similarly found that wind shear tended to drop when warm seas—in this case, those with temperatures above 83 degrees Fahrenheit—expanded their coverage of the tropical Atlantic.

Kerry Emanuel has been making the case for the link between rising air temperatures and rising sea-surface temperatures, but he, too, is quick to acknowledge that sea-surface temperature is only one of many factors affecting hurricane intensity. Over lunch in 2006 at the Tropical Cyclones and Climate workshop in Palisades, New York, Emanuel made a point of explaining to me that the rise in hurricanes' potential intensity with sea-surface temperature occurs in part because both have links to other factors. High winds cool the sea, and they often nip would-be hurricanes in the bud. Slower seasonal winds allow hurricanes to gather steam and also allow the surface to retain more heat. Volcanic eruptions, too, generally cool seas even as they produce other conditions that can thwart hurricane formation. "Part of the increase we see since the early 1990s is there haven't been many volcanoes," Emanuel explained, noting that the eruptions of Mexico's El Chichon in 1982 and the Philippines' Mount Pinatubo in 1991 almost certainly helped lower the intensity of Atlantic hurricanes during the earlier period of the record he analyzed. Volcanic particles tend to block some of the sun from reaching Earth's surface, which helps keep the sea surface cooler. Partly as a result, these particles tend to raise the temperature where they reside, in the upper troposphere. In contrast, hurricanes tend to be particularly powerful when a cooler upper troposphere sits above a warm surface—a common situation in recent years. "This is a big player, with what we've seen," Emanuel said.

Like any climatic factor, sea-surface temperature and other factors affecting hurricane formation sway in tune to natural variability as well as the longer-term influence of global warming. Still, even those who don't believe in the connection between global warming and stronger hurricanes have cause for concern based on the records of the past gathered by Kam-Biu Liu of Louisiana State University. He has looked at sediment cores from coastal lakes to tally dune-topping storm surges. "Each record would give us a sand layer, or fingerprint if you will, of a past storm event," he said during the 2006 Tropical Cyclones and Climate workshop. He has found a hyperactive period of storm activity stretching from about 3,800 to 1,000 years ago based on four cores from four different U.S. coastal states. Based on Liu's findings, even the past thousand years have been relatively mild. "We really haven't seen anything yet. What we've seen in the past, we may see in the future," he commented. After describing his work for U.S. coastal areas as well as some coastal regions of China, he urged caution when considering how future hurricanes might strike. "If the officials are basing their assessments on the last 150 years, then they are missing the boat," he said. "And they may also be putting a lot of lives at risk."

Liu's concern received support from a later study of coastal lake sediments linked to landfalling Atlantic hurricanes, described in a 2009 *Nature* paper by Michael Mann and colleagues. They compared these overwash sediments for the past 1,500 years collected from sites in New England, the mid-Atlantic, the southeastern U.S. coast, the Gulf Coast, and the Caribbean. Their results showed that the frequency of these dune-topping storm surges peaked in medieval times (especially from about 1000 to 1200), then subsequently dropped to lower levels of activity for about five centuries (from about 1250 to 1750). Their independent comparison using computer modeling of other data suggested that during the peak, tropical Atlantic sea-surface temperatures were running high, even as the atmosphere favored a climate pattern that promotes lower wind shear in the Atlantic (that is, La Niña-like conditions in the Pacific Ocean). During the subsequent hurricane lull, their model suggested that these two situations reversed, with lower sea-surface temperatures and conditions ripe for higher wind shear. While the authors don't highlight this fact, the peak they found occurred during the time frame known as the Medieval Warm Period. Meanwhile, the decline extended across much of the time frame encompassing the so-called Little Ice Age.

❧

Limited evidence suggests that hurricanes weakened as well during the last full-blown ice age and strengthened beyond anything we've seen yet during hothouse climates. First a caveat: It's difficult to assess global hurricane activity during past climates. Sediments can reveal information about particular regions. Researcher Kam-Biu Liu of Louisiana State University and others have used sediments to begin piecing together some evidence for hurricane activity in specific coastal areas in the United States and China. But these records extend back only a few thousand years—an admirable accomplishment, but falling short by a long shot of the termination of the last ice age roughly 10,000 years ago. Also, hurricanes may strengthen in some regions while weakening in others, making a global assessment out of reach without a vast increase in research covering more space and time. For the moment, at least, researchers have to resort to computer models to consider how hurricane regimes might have differed in the past.

Jay Hobgood and Randall Cerveny tuned a global climate model to conditions of the coldest depths of the past ice age to see how hurricanes might fare. They concluded in a 1988 paper that these storms could still operate in this glacial climate, but in a weaker state than during interglacial warm periods, including the current climate. In our modern climate, hurricanes won't form unless tropical sea-surface temperatures reach about 80 degrees Fahrenheit. The threshold temperature for hurricane formation would have dropped slightly in the ice-age atmosphere, just as it is likely to rise somewhat in warmer times. Even taking these changes into consideration, hurricane intensity likely would decline in ice-age climates, these researchers concluded.

The situation would reverse in hothouse climates. In fact, the evolving understanding of how hurricanes function under different climates might help explain a long-standing puzzle about earlier hothouse worlds. Oceanographers and climatologists have struggled to explain two irreconcilable "facts" about earlier hothouses such as the Eocene and Cretaceous—one, a warmer global climate, with little or no permanent ice even at the poles, and two, tropical seas that would have made surfers reach for their cold-water gear. How could sea-surface temperatures bordering the equator run between 59 and 73 degrees Fahrenheit, as the oxygen isotope evidence suggested, when palm trees were swaying at the latitude of modern-day Chicago? This "cool tropics paradox" caused climatologists to scratch their heads and computer models to sputter. Climate models needed tropical sea-surface temperatures above 86 degrees Fahrenheit in order to reproduce the warm polar Eocene climates that came out loud and clear in the fossil evidence of the animals and plants that lived at these high latitudes.

In the past decade, evidence arrived from several fronts to challenge the illogical results showing that ancient hothouses featured cool tropical seas. Paul Pearson of Cardiff University and colleagues suspected that the material used in earlier research had been too old and crusty to provide reliable dates. Sitting on the cold ocean floor for 50 million years (Eocene) or 100 million years (Cretaceous) might alter the information locked in the shells, they reasoned. So they sought out pristine samples from materials that had retained their shape and glassy look after all those years, tossing out any specimens that had begun to blur around the edges. Results published in 2001 from these pristine samples showed that tropical sea-surface temperatures averaged between 82 and 90 degrees Fahrenheit during the late Cretaceous and Eocene hothouses. A follow-up effort in 2007 put the tropical sea-surface temperatures around Tanzania between 86 and 93 degrees Fahrenheit throughout the Eocene. The high end is several degrees hotter than is reached in the open ocean today and about the temperature we're starting to get in summer in some parts of the Gulf of Mexico and Caribbean Sea.

"It's a useful rule that the sea-surface temperature never exceeds 86 degrees Fahrenheit in the open ocean today—it only happens in isolated basins," Pearson explained in an e-mail exchange. "So, yes, the temperatures are looking quite toasty." Other researchers used a similar approach when reexamining tropical sea-surface temperatures during the hothouse Cretaceous. Karen Bice of the Woods Hole Oceanographic Institute headed a research effort to examine pristine material from the Cretaceous hothouse. Their results, published in 2006, suggested that sea-surface temperatures in the tropical Atlantic around Suriname reached 91 to 95 degrees Fahrenheit during much of the Cretaceous. Now that's getting warm—but perhaps still tolerable with a proper sea breeze.

Sea temperatures that high can spur on more than a sea breeze, of course. There is some evidence that high sea-surface temperatures during past hothouses provoked storms of an intensity far beyond what we've seen in the historic record—even beyond what sediments have recorded for the past several million years. In a creative approach, Japanese researchers used lines in the sand to compare storm intensities over the past quarter of a billion years. Makoto Ito and his colleagues at Japan's Chiba University measured the distance between wavelengths preserved in ancient sandstone that fit the profile of near-shore storm deposits—pockets of fine-grained sand of the sort pulled into the sea during hurricanes and other intense storms. These "tempestites," named after the tempests that create them, differ in appearance from the silty muds that settle to the seafloor in calmer times, sandwiching them in. The color and texture of tempestites often stand out, making visible the curvy imprint of the wave action they preserve in hummocks as sand begets sandstone. These hummocky formations originate from waves washing over the seafloor, rearranging sand into a series of raised bumps separated by slight depressions. Like washboard bumps on a dirt road, the bumps in hummocky tempestites tend to be evenly spaced. The "wavelength"—that is, the distance between the bumps—reflects the severity of the storm, based on analyses of modern tempestites. Basically, bigger waves yield bigger wavelengths.

Ito and colleagues measured the wavelengths of hundreds of storm deposits around Japanese islands, with samples dating from about 260 million years ago during the late Permian through a million years ago during the Quaternary. The Cretaceous, which stretched from 144 million to 65 million years ago, contained numerous samples. When they pulled their data and findings from similar studies together onto one timeline, published in 2001, they found that storm intensity peaked during the mid-Cretaceous—right about when global temperatures peaked in the half-billion-year record. The tempestites laid down during the mid-Cretaceous contained wavelengths about 20 feet apart, triple the distance of those formed by even the most severe modern storms.

Circumstantial evidence like this cannot prove causality. In this case, the researchers cannot even prove that hurricanes caused these formations, as they note. And the results do not reveal exactly how big those storms were, say, on the Saffir-Simpson scale of hurricane destructive power. Still, the concept that bigger storms would create greater wavelengths fits our understanding of how these formations work in modern times. And the greater wavelengths found during the hothouse Cretaceous supports our understanding that higher sea-surface temperatures can support more intense storms. It also supports the argument that we should do our best to avoid raising global temperatures high enough to bring the planet back to a hothouse state.

Factors that can dampen hurricane strength—such as wind shear, dusty skies, and cool ocean currents—may intervene along the way to keep wind speeds from escalating off the charts as the world warms. Even the hurricanes themselves help cool down the tropics. But we are entering uncharted territory on ocean temperatures, at the surface and below, so all projections contain elements of speculation. What we do know should give us cause for concern. We know hurricanes require warm sea-surface temperatures to get rolling and that warm waters at the ocean surface and below power their winds and rains. And we know global warming heats up the oceans, almost by definition. So coastal residents in particular and society in general better prepare for a continuing onslaught of strong hurricanes, typhoons, and tropical storms as long as we continue to warm the planet. Not only does the physical evidence show that these storms intensify as ocean temperatures rise, but Gaian logic suggests that there are good reasons for this reaction. So there is a silver lining to these cloudy skies. Hurricanes help Gaia survive and moderate global warming, even cooling tropical seas on a real-time basis. What's more, hurricanes can help cool things down via their effects in the biological realm.

❧

Hurricanes also seem to assist Gaia's efforts to beat the heat via plants, which draw down heat-trapping carbon dioxide. Research efforts by dozens of scientists point toward similar results: In both the forest and the sea, hurricanes boost productivity—meaning the hard work of plants as they build carbohydrates out of carbon dioxide, water, and sunlight. It's an important job, as atmospheric carbon dioxide shoulders the blame for about three-fifths of the modern global warming. By the start of this century, carbon dioxide levels were about a third higher than they were before the Industrial Revolution brought us cars, coal-powered electricity, and central heating. By the end of this century, levels could double or even quadruple preindustrial levels. Anything that helps pull carbon dioxide out of the air potentially helps slow global warming.

Plants pull carbon out of the air and add hydrogen from water molecules to construct carbohydrates—the building blocks of life. Carbohydrates form marine algae and forest leaves as well as longer-lived organic carbon structures, such as wood and soil organic matter. Hurricanes shift some of these constructs of carbon dioxide into potential deep storage when their floods carry leaves, wood, and soil organic matter into waterways, wetlands, and the sea. It may seem ironic that any living thing could benefit from this force of destruction called a hurricane. Maybe it's a matter of life celebrating its survival. More likely, it's life enjoying the boon of nutrients coming up from the seafloor, down to the forest floor, and from the river to the sea.

In the ocean, hurricanes boost the growth of marine plants, including algae and microscopic plankton. Like Robin Hood, a hurricane redistributes the wealth. It lifts valuable fertilizers from nutrient-rich deep waters and sediment deposits into the nutrient-poor surface waters where the masses live. In fact, the nutrient deficit at the sea surface specifically stems from the crowding of plant life into the top 300 feet. Known as the photic zone, it's the only place marine plants can catch some rays. In the churning wake of a hurricane, plankton and other plants bloom in numbers as they feast on nitrogen, phosphorus, and other micronutrients. Post-hurricane bursts of algal productivity show up in satellite images, under the eyes of those who know how to detect photosynthesizing chlorophyll. Researchers Amélie Davis and Xiao-Hai Yan found that chlorophyll levels typically doubled in the days after a hurricane's passage, based on their satellite-imagery study of seven storms passing over the northeastern U.S. continental shelf. Davis cautioned that the results should be taken with a grain of salt—or perhaps a grain of sand—because satellite imagery can confuse algae with sediment. Still, in one remarkable example, the Category 5 Hurricane Isabel produced a 340-mile-long strand of algae that drifted around the coastal area for two weeks before disappearing from their radar, as it joined the food chain or sank toward the ocean floor.

Admittedly, algal blooms are not the most desirable forms of life. Some might call them the scum of the earth. But at the microscopic level, they're just another host for chloroplasts, a member of the bacteria group that Gaia theory co-developer Lynn Margulis credits for much of the planet's regulatory ability (more on this in the next chapter). Algae draw down carbon dioxide as they create tissue. By using carbon dioxide from the water, algae help create a gradient that can pull more carbon dioxide from the air above. When their short lives end and any remaining algae start to rot, photosynthesis shifts into reverse. Decay sets in: The carbon in dead algae links with oxygen in the water to re-create carbon dioxide. The ongoing decay depletes local oxygen supplies. The dearth of oxygen in the waters around a bloom can suffocate or repel life forms that otherwise might dine on algae, living or dead. The lack of oxygen can limit decay as well. If even a small fraction of the algal bloom escapes decay to land on the seafloor, some of this carbon gets a sea burial—in some cases, alongside the carbon contributions from dead bottom dwellers or oxygen-starved fish. Not a pretty sight, but potentially a way for Gaia to draw down airborne carbon dioxide.

Along with stirring up plant fertilizer from deeper waters and seafloor sediments, hurricanes shift nutrient-rich sediments from land to sea for marine plants to exploit. For instance, Hurricane Gilbert's passage in 1988 pulled an estimated 92,000 tons of sediment into Jamaica's Hope River that September—twenty to thirty times the usual monthly quota based on September values for two subsequent years. Because hurricane rainfall tends to be intense yet long lasting, the floods it produces often carry a disproportionate amount of sediment and carbon compared to other rainfall events. Some of the sediment from mountains and valleys eventually settles down into deltas, wetlands, and the ocean.

Hurricane Katrina transported enough sediment to coat streets and the floors of buildings throughout flooded regions of Louisiana. I heard about some local manifestations of this when I stopped at a convenience store at the Exxon station in Leesville, a small southern Louisiana town on a tongue of land formed by Mississippi River sediments. In the course of laugh-filled conversation on a late Friday afternoon in December 2005, I mentioned to the group of locals chatting over cans of Budweiser that the people in Leesville seemed to be taking the storm with a good sense of humor. "Sure, now we are," said one man. It was tough to chuckle in the immediate aftermath, added Dodie Thomassie, a longtime employee who grew up in the area. "The water was five and a half feet in here," Thomassie said, pointing to where the waterline had reached on the wall. I looked around, marveling at the spotless white walls and clean shelves stocked with fresh provisions like dish soap, tuna, and soup. "And we had a four-inch layer of stinky mud on the floor," she added. Hurricane Katrina also deposited an eight-inch layer of mud on the Mississippi town of Hancock (near Biloxi), according to a September 7, 2005, *New York Times* story. Sediment layers on paved streets and in buildings are easier to measure, but comparable deposition layers also make their way to mangroves, bayous, and coastal waters. Even more sediment used to arrive before the levees and dams intercepted the flow to the sea. While annoying to building owners, these sediments and the nutrients that come with them have helped make the Louisiana bayou one of the world's most productive ecosystems.

This outpouring of sediment can leave a hurricane signature on the seafloor, especially in coastal regions. Studying sediment cores from the Everglades and Florida Bay, researchers Woo-Jun Kang and John Trefry found evidence for the passage of three hurricanes: Donna in 1960, an unnamed hurricane in 1948, and Hurricane Labor Day in 1935. All of that hurricane destruction and production left telltale signs on the ocean floor. For one thing, the hurricane sediment layers held roughly half the usual ratio of phosphorus to carbon, signaling it had been scavenged by life for phosphorus, a nutrient that promotes plant growth. For another thing, hurricane years left behind thicker sediment deposits than non-hurricane years. Kang and Trefry found hurricane layers as thick as 10 inches in the sediment record. That's pretty thick, considering scientists measure mid-oceanic sediment layers in millimeters or even micrometers. Finally, hurricane sediment deposits generally contained two distinct layers: a lighter one of shells and other marine-based calcium carbonates, topped by a darker layer of land-derived plant matter and soil. In the dark layers, Kang and Trefry found microscopic bits of root hairs from nearby mangrove forests.

Even the logs carried from forest to sea during hurricanes share nutrients to boost marine productivity. Naturally formed rafts of logs carried to the sea during storms typically float around the ocean for a year or so before they sink. During this time, they serve as a floating feast for an ecosystem of fish, including the ever-popular tuna. In fact, tunas' attraction to seaborne logs has inspired some fishers to launch their own log rafts as tuna-locating devices. Log rafts of any source can turn into carbon stores as well if they escape decay during their eventual journey to the seafloor. Not many creatures have the talent for turning wood into food demonstrated by termites. Further, lack of oxygen in the deep ocean and other factors limit the decay of waterlogged wood. That's why sunken ships can endure for hundreds of years at the bottom of the sea.

❧

Back on land, hurricanes similarly drop loads of nutrients onto the forest floor. Their winds knock down trees and branches, leaving them strewn across forest paths and city streets alike. The floods and landslides that come with hurricanes spirit off tree trunks and branches, sometimes depositing them under soil, in riverbeds, or on the ocean floor. While these logs languish, the trees that remain alive—and even many initially taken for dead—put on new leaves. At the same time, fast-growing pioneers shoot up in pockets of sunlight created by the downfall of other trees. For these reasons, forests struck by hurricanes can act as "carbon sinks," pulling carbon dioxide out of the air and into the pool of earthbound carbon. That's what research efforts at the International Institute of Tropical Forestry (IITF) in Puerto Rico have shown. Hurricanes create a carbon sink that lasts for at least decades, surmises longtime IITF director Ariel Lugo.

"The beauty of hurricanes is that they create fast-term regrowth, with ongoing storage. I think it's nature's way of readjustment," as Lugo put it during a telephone conversation in 2006. I first met Lugo in 1989, when he zoomed up on his motorcycle to help me get started on the IITF internship that brought me to Puerto Rico about a week before Hurricane Hugo. After the hurricane struck, it looked like the apocalypse. With all the barren trees, I almost expected to see smoke rising from the forest floor. Yet the "fragile rain forest" soon demonstrated a phoenix-like ability to rise from its ashes. This rapid response differs from the lingering destruction that can follow burning or clear-cutting of tropical forests.

Hugo downed about 44 percent of the trees in the hardest-hit section of the Caribbean National Forest. That's what Lugo and a host of other IITF researchers documented in a December 1991 special issue of

Biotropica about the hurricane's impact on the local rain forest, known as the Luquillo Experimental Forest in the scientific literature. (Here and throughout this book, I will use the term "rain forest" in the sense of the public vernacular to encompass both moist forests and wet forests and the term "tropical forests" when including dry forests as well.) With their branches and leaves shorn off, many remaining trees resembled telephone poles. Yet within seven weeks of Hugo's strike, all the living trees had put on a new set of leaves, ecologist Lawrence Walker reported. Over the next year, he found that about 80 percent of damaged trees had resprouted. Trees that had snapped in half sent out new shoots from their broken trunks, while many of their uprooted brethren converted some of their remaining branches into saplings. Similarly, researcher Katherine Yih and her colleagues found rampant resprouting among tropical trees in a million-acre section of the Nicaraguan rain forest struck by Hurricane Joan in 1988. Joan's 156-mile-per-hour winds had downed more than three-fourths of the trees in their research sites and stripped the leaves off most remaining trees. Four months later, about 83 percent of the tropical hardwoods had resprouted. Only a quarter of the trees, mainly pines, had failed to resprout. Apparently many tropical hardwood trees have evolved to take hurricanes in stride.

Back in Puerto Rico, the jungle's miraculous revival repeated itself after Hurricane Georges in 1998. Immediately after the storm, we had to chainsaw our way through the deadwood to get down from my friends' mountaintop home outside Luquillo. We could barely walk three strides without coming across another fallen tree or huge branch. My research across Puerto Rico over the next few months showed Georges had downed some 20 percent of island trees, although the damage ranged from 1 to 92 percent depending on species and location. Yet within four weeks, so many leaves had grown back that I had to abandon my plan to compare canopy leaf loss among plots. Although I didn't officially tally resprouting from Georges, many of the broken or fallen trees I encountered also showed renewed signs of life in subsequent months.

Cities as well as forests deal with loads of downed wood during hurricanes. When Georges swept through Puerto Rico, it left behind about 80 million cubic feet of debris in urban areas. Hurricane Katrina reportedly left behind about 1.6 billion cubic feet of debris in Louisiana alone. Along with debris from gutted homes, deadwood covered the streets of New Orleans. "There were so many trees down, you couldn't even drive in the city at first," recalled Joe Braun, a saxophonist with the Jazz Vipers who described his Katrina experience between sets in the French Quarter. "All the live oak trees lost branches. They've got branches as big as trees. They were all over the street," he remembered. Even four months later, some uprooted trees continued to litter the cityscape, including a live oak with a trunk the size of two refrigerators. It had gently landed on the house next door without doing any apparent damage. Despite all the downed urban trees, only two Katrina-related deaths resulted from falling trees, both in Florida. By December, many of the city's standing trees sported fresh foliage. Clearly the trees were doing their part to "Re-New Orleans," as the bumper stickers instructed.

The revival of barren and even fallen trees represents only one aspect of a burst of plant productivity that surges through forests after a major hurricane. In post-Hugo Puerto Rican forests, incredible growth rates continued for five years straight. Annual productivity rates were roughly triple the usual rates, as documented in a thorough study led by then-IITF researcher Frederick Scatena, one of my mentors and now chairman of the University of Pennsylvania's Earth and Environmental Science Department. He and his co-authors projected that the aboveground vegetation would reach its pre-hurricane biomass (that is, collective dry weight) within another two years if growth continued at the rate measured during the five-year study. These hurricane-impacted forests do tend to be smaller in stature than equatorial tropical forests that do not face hurricane winds on a regular basis. Still, modeling work by Robert Sanford Jr. and colleagues suggests that repeated hurricane hits increase the forests' overall productivity. In their model, this occurs largely because the fallen wood increases the amount of carbon stored in the soil (a topic of chapter 6).

At the bottom of landslides or other places where lack of oxygen limits decay, many kinds of wood can remain intact for centuries or even millennia. Decay requires oxygen, as mentioned earlier, and oxygen runs in short supply at the bottom of a 30-foot pile of soil. (The same holds true for most landfills, where the remnants of urban trees often meet an unfortunate end as wood chips.) Landslides abound during heavy rains, especially when those rains come with hurricane winds that uproot mountainside trees. Hurricane Hugo triggered a landslide for roughly every 10 acres in a 2,500-acre section of Puerto Rico's Caribbean National Forest surveyed by geologist Matt Larsen, working with lead author Scatena. The largest one moved about 1 million cubic feet of soil and debris into a nearby river. During Hurricane Mitch, about 200 million cubic feet of soil shifted in El Berinche landslide, taking down an entire neighborhood in Tegucigalpa, Honduras. Although they can usher in tragedy from the human perspective, landslides can create local pockets for carbon storage. And hurricanes create landslides.

❧

In short, hurricanes launch processes that play a role in cooling our planet. In the biological realm, hurricanes contribute to carbon sinks by burying logs, soil, and sediment in landslides, the ocean, and anywhere they can better escape decay. Hurricanes also heave logs onto the forest floor, where they can languish for decades—perhaps even centuries if they are large and decay resistant. While the fallen trees and branches decay over varying lengths of time, the natural hardwood forests quickly revive—often into one of the most productive phases of their existence. Marine plants, too, thrive in a hurricane's wake, albeit on a shorter time scale. Hurricanes make waves that churn up nutrients in deep water and coastal sediments. The rains and floods that come with hurricanes pull carbon-containing sediment from the land into the ocean, where it can nourish marine life or wind up in long-term storage on the ocean floor. For every molecule of carbon locked up in vegetation or buried in deep storage, there's one fewer molecule free to form heat-trapping carbon dioxide or methane. Physically, hurricanes cool the ocean surface as their winds stir chillier waters from below into the mix. Their rains also disperse heat. The many cooling features of hurricanes help explain why they can help Gaia resist global warming.

With all that, the time scales involved in hurricanes' cooling actions do not necessarily match human needs. Mainly, hurricanes cool the world's two warmest regions—the tropics and subtropics. In so doing, they can actually warm regions closer to the poles. Even in the tropics, the cooling wakes hurricanes leave behind often seem to warm back up in a matter of weeks under the sun's hot glare. Although it seems logical, by extension, that these storms cool tropical seas on longer time scales, the idea that they could help from one season to the next remains speculative at this point. Similarly, a hurricane's boost to plant growth makes only the tiniest dent in global carbon dioxide levels—even if you add the efforts of marine and land plants together. The amount of carbon dioxide a hurricane pulls down to earth represents less than a drop in the bucket considering how much exists in the atmosphere. But any process that pulls some of these molecules back down to earth could help moderate global warming over the long term.

Hurricanes could help balance carbon dioxide levels at the scale of perhaps decades, but more likely centuries. Yet people are putting more carbon dioxide and other heat-trapping gases into the air on a daily basis, throwing the balancing act out of whack. Every year, carbon dioxide levels climb by about half a percent, on average (and rising), over their existing levels. We'd all be running for the hills far from the coast if hurricanes were going to control this excess. More realistically, hurricanes could help restore balance once we have our fossil-fuel habit in check. Given the growing sentiment that we're launching an irreversible warming of Earth, it's comforting to know that Gaia has at least some defense mechanisms for dealing with the temperature rise, even if hurricanes lurk among them.

Hurricanes are both a symptom of global warming and one of its cures. It's a tough pill to swallow. We would prefer if the antidotes to rising temperatures came in easier-to-take forms, like the minuscule tablets of homeopathic medicine. But sometimes a body—even a planetary body—needs strong medicine. Hurricanes certainly pose a risk to the system, but they serve as one of Gaia's natural defenses to rising temperatures.

An informal poll of family and friends indicates that Gaia theory remains unfamiliar or incomplete to many Americans, and even some environmentalists. Among scientists, too, Gaia theory takes many different forms, including Earth system science. So, before delving into other ways the entity on which we all depend responds to changing temperature, it's time to get better acquainted with the theory that our living planet has a built-in system for climate control.

2

A Living System

Gaia and Climate Control

With the birth of Gaia theory in the mid-1960s, a new worldview began to emerge in Western science. Or rather, ancient ways reemerged, with a modern twist. In this holistic view, life-forms combined, congealed, and otherwise clumped together to form bigger systems. The holistic view came to embrace cells, ecosystems—even planets. The world was alive. Mother Earth had reincarnated as Gaia. Once the Greeks' vision of the Earth goddess, Gaia now had a decidedly space-age flair. While seeking evidence of life on other planets, British chemist James Lovelock had recognized her through the mists. He'd seen her face in an atmosphere that defied entropy. He identified the Earth—Gaia—as the world's largest life-form, a composite of smaller systems that emerged as a living system in its own right. At the other end of the spectrum, the American microbiologist Lynn Margulis had peered down at paramecia and other one-celled creatures and found them teeming with life, a universe unto themselves. The worldview captured by these two holistic approaches at opposite extremes also built upon the momentum of a growing understanding about medium-sized systems. Ecologists had already shown that life-forms cooperated in complex ways in rain forests, prairies, coral reefs, and other ecosystems. Still, applying this holistic view to the planet required vision and courage.

Lovelock's version of Gaia only loosely relates to the Greeks' vision of the Earth goddess. As with most of the gods and goddesses in their pantheon, Gaia's life in Greek myth resembled a soap opera more than a spiritual treatise. Born of Chaos, she gives birth to Uranus, identified variously as the universe and the sky. With Uranus's fertile rains, she gives birth to the mountains and seas and other earthly creations, including a host of children. Some Greek concepts of Gaia had aspects in common with the modern view, though. As Bruce Scofield noted in *Scientists Debate Gaia*, the philosopher Thales of Miletus, who lived in the sixth century B.C., saw the world as an organism—specifically an animal. Thales and his contemporaries, including Pythagorus, embraced the concept of a living Earth they called *anima mundi*. Many cultures outside of Greece also featured the concept of the Earth as a living being, typically a bountiful female. Some indigenous American cultures have sacred teachings featuring both Mother Earth and Father Sky—an interesting overlap with the Greek's concept of Gaia and Uranus/Sky (without the Oedipal overtones). In Colombia and other South American countries, people honor the Earth goddess Pachamama. Celtics honor Danu. Pagans in general honor local versions of Mother Earth, as do Hindus and some other Asian cultures. Lovelock could have bypassed some of the spiritual connotations by naming his theory Biocybernetic Universal System Tendency/Homeostasis, which he had considered. At the suggestion of neighbor William Golding, author of *Lord of the Flies*, he settled on the more poetic and succinct Gaia in the early 1970s. It suited the colorful spirit of the times. Anyway, as Lovelock likes to point out, the goddess Gaia is also known as Ge—as in geology, geography, and what he sometimes calls the field of geophysiology.

The twentieth-century version of Gaia includes an important scientific refinement. Lovelock established that life plays an important role in moderating Earth's climate. Methods include manipulating atmospheric gases, changing the planet's ability to reflect sunlight, and releasing compounds that help seed raindrops. Gaia theory starts from the premise that the Earth's climate has remained relatively stable despite an ongoing increase in solar power. Like every star, the center of our solar system—Sol—starts out using hydrogen for power. Then it gradually turns to helium, which tends to burn hotter. After observing other suns, astronomers estimated that Sol has probably warmed by about a quarter since life first began on Gaia. What's more, the position of Earth in relation to the sun changes in regular cycles related to its orbit and tilt. These Milankovitch cycles, as they're known, affect how much solar energy reaches Earth's surface. They help explain the timing of ice ages and hothouses over the past 2 million years or so (as the next chapter will explain). Despite the sun's ongoing warming and orbital fluctuations in the amount of solar energy reaching Earth, Gaia manages to maintain its temperature within a range suitable for life. Lovelock dubbed this planetary skill "homeostasis," thus comparing it to our own unconscious ability to maintain a stable body temperature.

Margulis prefers the term "homeorrhesis," as she mentioned during an aside at a 2006 conference. As she noted, Earth actually has numerous possible "body temperatures"—that is, relatively stable climate states. Lovelock also moved toward this concept in his 2006 book, *The Revenge of Gaia*. He compared Earth to a camel, noting that a camel has two relatively stable body temperatures: about 104 degrees Fahrenheit during the heat of the day, and 93 degrees Fahrenheit during the cool desert nights. Either way, the camel's temperature rises and falls within a predictable range.

This model of a living system that can survive within a specific range of internal temperatures is useful when contemplating past climate changes. Judging from ice-core records of Earth's temperature for the past 800,000 years, the planet can stabilize around at least two very different thresholds—the cold of ice ages, and the warmth of interglacial periods. At this time scale, and probably for much of the past 2 million years, Gaia often shifted from cold to hot fairly suddenly, with little pause between the two states. Such steplike behavior suggests a "quantum" phase change, such as when an H_2O molecule changes from liquid water to gaseous water vapor. It takes a specific quantity of energy to make this shift into another state, which is where the "quantum" comes in. Once the amount of energy arrives to make the change—that is, once it reaches that threshold—change can occur rather abruptly. If Earth behaves in a quantum fashion, it may be acting more like a molecule than a camel.

Interestingly, some support for the quantum concept regarding planetary climate changes comes from the realm of physics. Kerry Emanuel, a well-respected physicist at the Massachusetts Institute of Technology, devised a simple physical model to consider how the planet's greenhouse gases and circulation patterns might differ during ice ages, warm climates, and hothouse times. Using this simple approach with about a dozen formulas to describe the atmosphere and oceans in a 2002 paper, Emanuel found that his modeled planet could leap virtually from one state to another and remain stable there. Like the water that covers 70 percent of its surface, Earth could be stable in a relatively frozen state (an ice age), an intermediate state (such as our modern interglacial), or a hothouse state (with extra water vapor and other greenhouse gases filling the air). This may not be good news for Earth's inhabitants, as we still can't predict at what point the planet would reach a quantum threshold.

Emanuel's model focused on Earth's physical aspects, without invoking the role of life in these transitions. With life at the helm, the planet might be acting as a giant cell, using the metaphor Lewis Thomas suggested in his 1974 book, *The Lives of a Cell.* Although he never called Gaia by name, Thomas clearly was alluding to Lovelock's ideas in the book's namesake essay when he said, "I have been trying to think of the earth as an organism, but it is no go. . . . It is too big, too complex, with too many working parts lacking visible connections." In a burst of insight, he envisioned Earth as a cell. Continuing this provocative analogy in the book's final essay, he compares the planet's atmosphere to a cell membrane:

> It has the organized, self-contained look of a live creature, full of information, marvelously skilled in handling the sun. It takes a membrane to make sense of disorder in biology. You have to be able to catch energy and hold it, storing precisely the needed amount and releasing it in measured shares. A cell does this, and so do the organelles inside. . . . When the earth came alive it began constructing its own membrane, for the general purpose of editing the sun.

Thomas's book also described some of Lynn Margulis's groundbreaking ideas about cells. Margulis, after countless hours of peering through microscopes in a University of California–Berkeley graduate program in microbiology, became convinced that paramecia and other nucleus-bearing cells represented a community of life-forms. Her 1963 Ph.D. dissertation and subsequent work revived a long-discarded theory that cells were composed of several different types of bacteria working together as a unit. She surmised that these bacteria had given up independent existences to serve as "organelles" (akin to our lungs and other organs) for this larger being. One bacterium earned its keep by using solar energy to make food (chloroplasts). Another took up residence converting food back into energy (mitochrondria). Her work on this cell symbiosis theory was greeted with derision, then grudging acceptance, then distinguished awards. Now it is textbook material. DNA analyses later proved that the chloroplasts and mitochondria within cells do indeed have distinct genetic origins that differ from the cell's nucleus. It's been more challenging to find the evidence to support the final leg of her theory—that independent spirochetes latched on as flagellates and cilia to give cells mobility. Still, the success of at least two-thirds of her hypothesis eventually won her membership in America's prestigious National Academy of Sciences.

Given her early and unflagging recognition that different life-forms can work symbiotically, Margulis instantly recognized the value of Gaia theory when Lovelock presented it as a fledgling hypothesis at a scientific meeting in 1969. Lovelock recalled that she was one of only two scientists at the meeting to embrace the concept. Margulis soon joined forces with Lovelock to write two scientific papers, both published in 1974, with the two alternating as principal author. Lovelock described her role as "adding substance to the wraith of Gaia." One of their papers went into the journal *Tellus*, which appropriately translates as the Roman version of Gaia. The other ran in *Icarus*, a journal then edited by Margulis's husband at the time, Carl Sagan, also a NASA colleague of Lovelock's. A couple of years earlier, Sagan and a colleague had described in the prestigious journal *Science* their own take on how planetary gases influence the climates of Earth and Mars. Reportedly, Sagan appreciated Lovelock's space-age version of Gaia, although he didn't fully agree with the theory. In following decades, Margulis championed the Gaia concept throughout the United States, while Lovelock worked on the United Kingdom. Lovelock's 1979 book, *Gaia: A New Look at Life on Earth*, carried the concept worldwide.

The holistic approach applied by Margulis at the cellular level and Lovelock at the planetary level had already taken hold in the field of ecology. Brothers Eugene and Howard Odum, American ecologists, advocated taking what they called a "whole-before-the-parts" approach to studying how ecosystems function. They embraced and expanded upon the concepts advanced in 1926 by the Russian scientist Vladimir Vernadsky that the plants, fungi, microbes, and animals in the forest functioned together as a unified whole with a unified goal—to keep conditions optimal for their existence. Howard Odum's work in the Puerto Rican rain forest helped inspire his views on how ecosystems use energy to help create conditions suitable for their existence. A 1959 ecology textbook by Eugene Odum in collaboration with Howard presages Gaia theory when discussing how ecosystems function, as the following passage shows:

> Although everyone realizes that the abiotic environment ("physical factors") controls the activities of organisms, it is not always realized that organisms influence and control the abiotic environment in many ways. A South Pacific coral island is a striking example of how organisms influence their abiotic environment. From simple raw materials of the sea, whole islands are built as the result of the activities of animals (corals, etc.) and plants. The very composition of

> our atmosphere is controlled by organisms. Indeed, without living organisms our world probably would be a relatively unchanging mass composed of fewer kinds of materials. Like our moon, it would be a dull world indeed.

Vernadsky had similarly contrasted Earth and the moon in describing the functioning of the biosphere—that is, the interdependent web of life that lives on and near Earth's surface. The Odums also credit him for being among the earlier scientists pondering how the biosphere divides into ecosystems (roughly, groups of organisms that function with non-living materials as recognizable units in time and place).

Although Lovelock had not read Vernadsky's work when first envisioning Gaia theory in the mid-1960s, he later acknowledged the views of Vernadsky and other predecessors. Lovelock, too, marveled at how Earth's liveliness distinguished it from its planetary neighbors. That was his job, in a way. He was working on a NASA project to help predict whether life existed on Mars. The project inspired him to consider how to detect life on other planets, preferably from afar, in whatever alien form that life might take. What makes Earth different from the other planets? He pondered that carbon dioxide dominates Venus's atmosphere, but exists only in trace amounts on Earth. The atmosphere of Mars strongly resembles that of Venus—lots of carbon dioxide, only trace amounts of oxygen. Mars's atmosphere is simply sparser, like a few wisps covering a bald planet. Venus's thick, flowing air, meanwhile, makes it easier to accept this universal fact: Air is a fluid. Venus and Mars have the atmospheres you'd expect, given how gases stabilize. In contrast, Earth's atmosphere features mostly nitrogen and oxygen. Lovelock recognized this composition as a signature of life because it goes against the laws of entropy.

It takes life to move against the tendency for structure to decay into simpler forms, known as entropy. "Entropy is the general trend of the universe toward death and disorder," as mathematician J. R. Newman put it. Anyone who works to keep a home clean encounters entropy, also loosely known as the second law of thermodynamics. Energy dissipates. Cleanliness crumbles to dust. But life moves against this current. We do our dishes, sweep the floor. Like the Greeks before him, Lovelock noticed that Gaia created order out of chaos. The Odums, too, described how life moves matter in the opposite direction of entropy. At cellular levels, life builds structure out of molecules, cracking some to release energy, combining others to create leaves, bone, and tissue. Without life's constant work, molecules like methane would quickly disappear from the atmosphere, decaying into other gases—mainly carbon dioxide and water, in methane's case. Without life on Earth releasing oxygen from its bonds, carbon dioxide would soon dominate our atmosphere, as it does on neighboring planets. In a metaphorical way, carbon dioxide molecules act as the biosphere's dirty dishes. If some life-form isn't hovering over the sink on a daily basis, they just pile up and spoil the atmosphere.

That's what happened at one point to Biosphere 2, the giant terrarium erected in the Arizona desert in the 1990s by some adventurers seeking to create a self-sustaining system. Inspired by Vernadsky's concept of the interdependent biosphere as well as Gaia theory, the international group was seeking to encapsulate a life-sustaining biosphere for potential export to other planets. The 3.15-acre system under glass started out with a typical Earthly atmosphere, about 21 percent oxygen, 78 percent nitrogen, and only trace amounts of carbon dioxide. A couple of years after its original directors sealed it as airtight as a spaceship, oxygen levels dropped to about 14 percent while carbon dioxide levels climbed. It turned out the rich soil inside the sealed environment had been releasing carbon, which ensnared oxygen molecules to create carbon dioxide. Much of the carbon dioxide, in turn, was being absorbed by the concrete structures within the complex. The local plant life was too sparse to balance out the ongoing loss of oxygen, so the atmosphere changed. With so little oxygen, the people voluntarily locked inside Biosphere 2 could barely rouse themselves out of bed, much less grow the food they needed to survive their self-imposed isolation.

On our planet (dubbed Biosphere 1 by this group), we don't have to worry about running out of oxygen. Plentiful plant life and bacteria keep the oxygen assembly line going. Soils as rich in carbon as those in Biosphere 2 exist in few earthly places (perhaps an Illinois farm after a large dose of manure). Even the carbon now locked in Arctic permafrost won't deplete oxygen supplies when it thaws out enough to escape its icy confines. Our atmosphere contains about 0.039 percent carbon dioxide, so even a quadrupling of levels would barely put a dent in our oxygen supplies, which run at 21 percent. What we have to worry about is a buildup in the trace amounts of carbon dioxide because of its well-established role in global warming.

❧

Plants and bacteria help draw down heat-trapping carbon dioxide, keeping the carbon and releasing the oxygen. Credit for creating the original high-oxygen atmosphere of Earth goes to bacteria—namely cyanobacteria, also known as blue-green algae—according to Gaia theory as fleshed out by Margulis. Bacteria continue to play a crucial role today, including as chloroplasts in photosynthesizing plants. As every schoolchild learns, plants take in carbon dioxide and release oxygen. Yet Margulis's and Lovelock's concept that bacteria and plants cleared the planet's air of most of its carbon dioxide and replaced it with oxygen was radical in the mid-1970s. At that time, textbooks ascribed our planet's abundant oxygen to the breakdown of water vapor (H_2O) followed by the escape of hydrogen into space, leaving behind oxygen. Geologists resisted the idea of rewriting their textbooks to accommodate Gaia theory.

Geologists weren't the only scientists turning up their noses at the concept of Earth as symbiotic unit, especially one named after a goddess. True, Lovelock's 1979 book on Gaia proved popular with the general public and remains on the shelves today. Yes, the Gaia concept quickly found its way into writing and discussions among New Age thinkers and other spiritual progressives. But scientists initially rejected, ridiculed, and ignored it. Some joked about a planet regulating its temperature by a committee of life-forms. How would it ever get anything done?

Temperature regulation remains a mystery for many life-forms, even ones that aren't planet sized, as Margulis noted in *Symbiotic Planet* (1998): "Mammals, tuna, skunk cabbage plants, and beehives all regulate their temperatures to within a few degrees. How do plant cells or hive-dwelling bees 'know' how to maintain temperature? Whatever the answer in principle, the tuna, skunk cabbage, bees and mouse cells display the same sort of physiological regulation that prevails across the planet."

Termites could join this thought-provoking list, judging from a presentation that biologist J. Scott Turner gave at a 2006 Gaia theory conference in Virginia. Turner and his colleagues studied African termite nest construction by making a plaster cast of one of the towering structures. They found that its drab, lumpy look belied a surprisingly sophisticated design filled with intersecting corridors that help keep interior temperatures comfortable. It turns out these minute creatures located entryways to pull in a midday breeze, even taking into account the tilt and location of the nest itself. What's more, all termite nests in the area possessed identical tilts and orientations. The result? A design worthy of a sustainable-architecture conference. But termites don't hire architects to create blueprints. These design skills that work so well in their hot, arid environment evolved over time, presumably by trial and error. Turner sheds light on how such designs relate to natural selection in his 2006 book, *The Tinkerer's Accomplice*. The evolution of what he calls "environmental physiology" is awe inspiring in its complexity.

For that matter, is the regulation of any system fully understood? Even the human body, arguably the most thoroughly studied homeostatic system, remains mysterious in many ways. We know that our bodies do regulate temperature and otherwise maintain themselves. Yet if we had to consciously and continuously make decisions about how to maintain homeostasis, we would soon be overwhelmed. As medical doctor and inspirational writer Deepak Chopra reminds in a 1993 book, *Ageless Body, Timeless Mind:* "A hundred things you pay no attention to—breathing, digesting, growing new cells, repairing damaged old ones, purifying toxins, preserving hormonal balance, converting stored energy from fat to blood sugar, dilating the pupils of the eyes, raising and lowering blood pressure, maintaining steady body temperature, balancing as you walk, shunting blood to and from the muscle groups that are doing the most work, and sensing movements and sounds in the surrounding environment—continue ceaselessly."

The human body, of course, has had millions of years of evolution to reach its current complexity. Presumably any of our predecessors who failed to produce the correct mix of internal chemistry didn't live to tell the story, nor were they likely to win the competition for a spot in the gene pool. In contrast, a planet evolving on its own—a population of one, in effect—seemed to fall outside the realm of natural selection. This concern fueled another line of attack on Gaia theory. How could Nature, red in tooth and claw by some accounts, weed out the unfit planet? For the survival-of-the-fittest concept of natural selection to work, Nature needed fodder from which to choose. To address these criticisms, Lovelock and Andrew Watson developed a computer model invoking natural selection as the means for planetary temperature control. The model also demonstrated that homeostasis could proceed without conscious decision-making by a planetary body.

Lovelock and Watson's initial model, described in a 1982 paper, featured an imaginary planet called Daisyworld, populated only with two types of daisies—a white variety, and a black variety. The researchers provided abundant virtual moisture so temperature alone would influence the survival rate of the two varieties. The results of the daisies' competition, in turn, would influence the planet's temperature. In their model, the influence involved how the two colors reflect sunshine. Humans have an intuitive understanding of this reflectivity issue. We favor white during summer and darker colors during winter. That's because white excels at reflecting the sun's rays, whereas darker colors are better at absorbing it. In the computer model, the daisies were programmed to respond as they might to natural selection: Black daisies absorbed more heat and so thrived in cool temperatures, while white daisies flourished in warmer temperatures. The survival of individual daisies reflected prevailing climatic conditions. But the selection at the individual level changed the *planet's* reflectivity. A planet cloaked in black daisies absorbed more heat, thus keeping cool years warmer than they might otherwise be. A planet shielded with white daisies reflected more sunlight, thus keeping things cooler as the planet's sun warmed up. Lovelock got similar results in following up with more complex computer models, as did others.

Daisyworld and subsequent models helped win over many scientists to Gaia theory. Time, too, opened minds. By the turn of the millennium, scientists generally had embraced the main concepts of Gaia theory, albeit under the guise of Earth system science. At a 2001 international scientific conference in Amsterdam, more than a thousand delegates signed a declaration that "the Earth System behaves as a single, self-regulating system comprised of physical, chemical, biological and human components." This is not too far from Lovelock's words in 1979: "The biosphere is a self-regulating entity with the capacity to keep our planet healthy by controlling the chemical and physical environment." In the 1980s, Lovelock had corrected the theory to include physical forces, specifying that life and its environment worked as a coupled system. "Self-regulating system" in the declaration suggests acceptance among these scientists of the homeostasis aspect of Gaia theory.

Homeostasis appears to fall in the middle of the five categories that James Kirchner outlined in a defining 1991 essay pointing out different aspects of Gaia theory. It's listed third, sandwiched between four other hypotheses set on a gradient of "weak" to "strong" Gaia. The more universally acceptable "Weak Gaia" hypotheses include "Influential Gaia," that the biosphere has a "substantial influence over certain aspects of its environment, such as the temperature and composition of the atmosphere," and "Co-evolutionary Gaia," which asserts that the biosphere and the environment evolve as a coupled system. The more controversial "Strong Gaia" includes "Teleological Gaia," which implies some intention by the biosphere in keeping the atmosphere in balance, and "Optimizing Gaia," which holds that the biosphere's manipulations are efforts to create or maintain favorable conditions for existing life-forms. In practice, homeostasis tends to be lumped into "Strong Gaia," as some Earth systems scientists embrace only the two weak aspects of Gaia theory.

❧

While it took time for Gaia theory to gain acceptance among Western scientists, indigenous scholars welcomed it from the start. Many indigenous philosophies express humans' interdependence with other species and the planet some call Mother Earth. "When I first read *Gaia*, I thought, 'Wow, the Western people are finally waking up,'" Lloyd Pinkham, a member of the Yakama Nation in the Pacific Northwest, said during a talk at a 2006 conference on Gaia theory in Arlington, Virginia. Vine Deloria Jr., a Standing Rock Sioux and an important literary voice for American Indians from the publication of his 1969 book, *Custer Died for Your Sins*, until his death in 2005, also viewed Gaia theory as an opportunity to think more holistically. In a 2001 book, *Power and Place*, he wrote: "The Gaia Hypothesis, among other new theories, suggests that we should begin to look at things organically and that we might indeed be a minor episode in a larger scheme of life. Whether this hypothesis proves fruitful enough to become a dominant paradigm in the social/scientific future is beside the point. The issue today is that we are no longer bound to use mechanistic models exclusively to tell us how to think about the world." His co-author of the 2001 book, Daniel Wildcat, a Yuchi member of the Muscogee Nation of Oklahoma and director and faculty member at Haskell Indian Nation University, wrote:

> Compare the scientific view to widely shared tribal views in which humans understand themselves to be but one small part of an immense complex living system, something like Lovelock's Gaia Hypothesis. This hypothesis offers a holistic worldview in the most profound sense, where attention to relations and processes is much more important, at least initially, than attention to the parts of our experience. The point should be obvious: we, human beings, in all our rich diversity, are intimately connected and related to, in fact dependent on, other living beings, land, air, and water of the earth's biosphere.

Gregory Cajete, a Tewa Indian and author of *Native Science*, acknowledges Gaia theory as the closest concept in Western science to the Native American paradigm for viewing the world. "Today these ancient concepts are being reintroduced in the theories and philosophies guiding curricula on ecology and environmental systems, from kindergarten to university level instruction. The Earth is being perceived as a living, breathing entity—a megabeing if you will. These concepts generate still others, such as ecological relationships, ecological sustainability, and environmental stewardship; that is, the need to avoid polluting the land in which one lives, to care for the land."

❦

Caring for the land can include letting it respond to the ongoing climate change. Putting this concept into the framework of Lovelock and Watson's Daisyworld, individual species and the ecosystems they form act as the black versus white daisies. They are subject to natural selection—yet they can instigate as well as respond to changes in climate. On Earth, different systems thrive under different conditions. In deserts, where moisture is the most precious commodity, succulent plants and cacti take root. In moist climates, hardwood trees grow where sunshine abounds, and softwoods fill in where cold winters or poor soils restrict the growth of hardwoods. In saturated soils, cattails, buttressed trees, and other types of wetland plants take up residence. Where the climate provides more moisture than needed to sustain a desert but not enough to keep the fires at bay during the warm season, grasslands and prairies often dominate. Woodlands interspersed with meadows may also thrive. Changes in species and ecosystems—and in the global proportion of forests versus grasslands, for example—also influence climate. (More on that in chapter 5.) The tapestry of earthbound life weaves a more complex and colorful mesh than the black-and-white Daisyworld, but the underlying principle remains the same. The plants and ecosystems covering Gaia's surface play a vital role in its health—and its climate control.

At this moment in time, pollutants, including the greenhouse-gas emissions behind global warming, threaten Gaia's health. In some views, the ongoing global warming amounts to a planetary fever. Anyone who has had a fever—whether associated with diarrhea, a sore throat, an infection, or a specific disease—will recognize that pain and suffering come with the condition. A fever signals trouble. Many of these fever-causing illnesses will fade away with some tender loving care—along with antibiotics and/or herbal tea. Yet anything that is serious enough to provoke a fever could worsen if proper steps aren't taken to get healthy again. In the United States, we might think of diarrhea as an inconvenience, or a sign of flu. But on the global scale, millions of people die from the dehydration of diarrhea every year. In the 1970s, diarrhea killed 4.6 million children under the age of five in a typical year, making it the leading cause of child mortality around the globe at that time. Yet a rehydration therapy introduced in 1979 helped bring the deaths from early-childhood diarrhea down to 1.5 million a year by 2000. Similarly, curing a dangerous fever, even a planetary fever, may depend more on natural remedies than high-tech fixes.

Most medicines have their roots in nature, even if they've been processed into pills for mass distribution. The original antibiotic, penicillin, grew from a mold. A variety of medicinal herbs, made into teas, tinctures, or pills, help some conditions that might otherwise lead to fever. Malaria can be warded off with fever bark—or the compound it yields, quinine. Even water can work wonders when used under the right circumstances, as in the rehydration therapy that helped so many children survive potentially fatal bouts of diarrhea.

What's more, fever itself can act as a means of ridding the body of unwanted invaders. Although it's still a bit mysterious, fever apparently has enough benefits that even some ectothermic ("cold-blooded") lizards will sun themselves into unusually high body temperatures when injected with certain unwanted bacteria. So, even though fever shows up as a symptom of trouble, it can also help effect a cure. Other physical reactions we find unpleasant sometimes restore health as well. There's nothing like a case of nonlethal food poisoning to bring home that point. A miserable situation, but the body's eviction methods generally do the trick.

This is where we should get worried about how humans fit into Gaia's system. Humans may be to Gaia as bacteria are to humans. Our internal bacteria comprise at least 10 percent of our mass, by Margulis's estimation in her 2009 book with Michael Chapman. The collective weight of humans to Gaia's mass pales by comparison—but our cities, farms, and ranches cover a good chunk of the land surface in place of the biosphere. So our impact on the planetary body we call Earth may be comparable to how some bacteria can affect our bodies. Bacteria can help our bodies to thrive. Serving as mitochondria, they power our activities by converting food compounds into energy. Some bacteria, such as acidophilus, aid digestion. Yet we can suffer serious illness when colonized by unfamiliar bacteria, such as the kind we might inadvertently ingest when traveling abroad among foreign flora. Modern humans might play a similarly health-threatening role from a Gaian perspective.

Our fate as a thriving civilization depends on Gaia's healthy recovery or eventual stability. And signs of trouble in our own bodies can enlighten our interpretation of Gaia's health, as philosopher David Abrams explains in an eloquent 1991 essay he published in *Scientists on Gaia*:

> When we consider the biosphere not as a machine but as an animate, self-sustaining entity, then it becomes apparent that everything we see, everything we hear, every experience of smelling and tasting and touching is informing our bodies regarding the internal state of this other, vaster physiology—the biosphere itself. . . . And this can be the case even when we are observing ourselves, noticing a headache that we feel or the commotion in our stomach caused by some contaminated water. For we ourselves are part of Gaia. If the biosphere that encompasses us is itself a coherent entity, then introspection, listening to our own bodies, can become a way of listening and attuning to the Earth.

If Gaia is feeling an internal upheaval from these excess greenhouse gases and the temperature rise they cause, it would behoove us all to listen hard. Attuning to Gaia's responses to global warming can help us understand what we can do to reduce the symptoms, or impacts. But first we need to consider how the modern warming is manifesting, and how this relates to previous excursions of carbon dioxide and temperature on the planet.

3

Greenhouse-Gas Attack

One Way to Warm a Planet

At the global scale, our impact on the planet may appear as invisible as the atmosphere itself at times. Viewed from a space perch, Gaia shines with her usual sapphire and emerald beauty during the day. But at night, all those sparkling diamonds of light signal we are here—and help explain why greenhouse gases are building up in the air. Most lights draw their power from fossil fuels, either directly by burning kerosene and gasoline or indirectly from electrical plants that run on coal, oil, and gas. At the local scale, we can sometimes see our influence with our own eyes, such as when smokestacks belch out pollution. Or we can smell it. In the mid-1970s, my family lived mere blocks away from a Chicagoland garbage dump for years before its operators started capturing the escaping methane, a potent greenhouse gas. On summer days, this natural gas would drift over with other fumes reeking of rotten fruit. Even people living far from garbage dumps and smokestacks can sense the presence of overabundant heat-trapping gases—almost every corner of the world has registered a local temperatures rise in recent decades.

Greenhouse-gas levels have been increasing with human productivity, which employs the burning of fossil fuels to get the job done. Electricity production from coal, oil, and gas—the infamous fossil fuels—emits carbon dioxide. So do gas-burning engines—planes, lawnmowers, boats, and fleets of cars. Americans drove 2.6 trillion miles in 2004, enough to get to Pluto and back nearly 500 times, an Environmental Defense study led by John DeCicco and colleagues showed. All this mobility and productivity has bumped up carbon dioxide levels in the air by a third since about the middle of the nineteenth century. Earth's temperature has responded in kind. Since what some consider a "climate shift" in the mid-1970s, the average planetary temperature has climbed upward, albeit with some wiggles. The wiggles capture the changing seasons, cooling events like volcanic eruptions, and year-to-year variability. (Variability guarantees that some years will be cooler than others even during a warming trend.) The climb reflects growing levels of carbon dioxide and other greenhouse gases in the air.

Carbon dioxide emissions account for about 60 percent of the modern temperature rise, while methane emissions from landfills, pipeline leaks, wetlands and agriculture account for about 20 percent. The rest of the current warming traces back to other pollutants, including nitrous oxide, ground-level ozone, and designer chemicals like chlorofluorocarbons (CFCs). This list describes the most common greenhouse gases, but just about any compound with more than two atoms can trap heat that might otherwise escape from the planet's atmosphere. Without any greenhouse gases, Earth's temperature would run below freezing. With too many of them, we have our current dilemma—temperatures rising, with no clear end in sight.

Just how high will it go? Air temperatures rose by about 1.4 degrees Fahrenheit from about 1900 through 2005, as the Intergovernmental Panel on Climate Change (IPCC) noted in 2007. While IPCC scientists project a "best-estimate" global temperature rise of another 3 to 7 degrees Fahrenheit by the end of this century, the range includes the outside chance of an increase of 11 degrees Fahrenheit. Some of this wide range relates to our uncertainty about the amount of greenhouse gases society will release into the atmosphere. The models consider different scenarios, ranging from a rapidly increasing world population with an economy powered by fossil fuels to a stable world population making a large-scale switch to alternative energy sources. But even with a known level of greenhouse gases—such as a doubling of carbon dioxide—the end results remain somewhat uncertain. Global climate models have improved over the decades, but the long-term temperature projections for a doubling of carbon dioxide remain much the same as in the 1980s—between 3 and 8 degrees Fahrenheit. The effects of clouds and other natural features, and even pollution itself, continue to limit the precision of projections for the future.

Don't let the model's imprecision about the exact number lull you into a sense that a warming itself is uncertain, though. The link between rising carbon dioxide levels and global warming becomes irrefutable when considered from time scales other than the last few decades. At the scale of minutes, the air temperature in a sealed flask filled with a carbon-dioxide-creating concoction of baking soda and vinegar will run higher than a flask of regular air if both receive the same amount of light. Similarly, explorations of ancient climates make it clear that our modern warming isn't the first time carbon dioxide and global temperature have risen in sync. These two have a long-standing relationship that has outlasted many a mountain.

Timelines spanning half a billion years exist for both temperature and greenhouse-gas levels, to varying degrees of precision. Beyond instrumental records, ice cores yield the most detailed records, which now span more than three-quarters of a million years. Even some details on these remain subject to debate, while records become less reliable beyond the ice cores' reach. Overall, though, the evidence drawn from deep time strengthens the link between high greenhouse-gas levels and high temperatures, as well as high sea levels. This chapter highlights those connections in modern and ancient climates, while subsequent chapters will elaborate on these and related changes from a Gaian perspective.

❦

Some skeptics who challenge global warming claim to be puzzled. They wonder how scientists can be so sure the planet is warming when in the late 1970s we were threatening the pending arrival of an ice age. Well, I was in high school at the time. But I remember hearing the ice-age-cometh rumors. After looking into this recently from the vantage of a scientist, I concluded it's wise to remember that newspaper headlines don't always reflect the scientific thinking of the day. The concern apparently followed the mid-1970s scientific breakthrough linking global ice-cover changes over the millennia to recurring changes in sun cycles (described in the next section). The researchers who made this discovery, including John Imbrie, pointed out that ice sheets had already been in retreat for more than 10,000 years, the typical span of an interglacial warm period. On top of that, the United States and much of the world had been experiencing a cooling phase for the previous 30 years. With help from a few scientists, the media converted this cooling and the breakthrough into a descent into an ice age. That's not what most scientists were saying, though. Even Imbrie's 1979 book, *Ice Ages: Solving the Mystery*, written with his daughter Katherine Palmer Imbrie, noted that the cooling registering since the 1940s was unlikely to last beyond a few more years. Society's input of greenhouse gases would reverse it at some point, the authors wrote. They even included a graph projecting a warming quite similar to the one documented as starting in full swing in the mid-1970s.

Later research indicated that much of the cooling related to the many pollutants in the air, especially sulfur dioxide and related sulfates. These are the pollutants that make smog so dangerous for asthmatics. Scientists call them "aerosols." More recently, this phenomenon has been dubbed "dimming," as in a 2006 NOVA documentary, *Dimming the Sun*. By whatever name, these sulfates literally block some of the sun from penetrating to the surface. Although the NOVA program portrayed the cooling effect of aerosols as a new discovery, scientists had understood for more than a decade that these aerosols were interrupting the warming effect of greenhouse gases. In its 1995 report, the Intergovernmental Panel on Climate Change used the 1991 eruption of Mount Pinatubo to estimate the cooling effect of aerosols. Like El Chichon in 1982, Pinatubo spewed sulfur compounds high into the atmosphere, putting them temporarily out of reach of raindrops and other elements that could pull them back down to earth. With the natural sunscreen lingering in the atmosphere, surface temperatures dropped and then plateaued for a year or so. They resumed their climb only after these sulfate aerosols had descended from their perch high in the sky.

Air pollution similarly has shielded us from the full effect of global warming. Sulfates released from smokestacks and tailpipes don't reach as high as those from exploding volcanoes, but every day we replenish the supply that gets rained out. Reducing air pollution—as many industrial countries began to do in the 1970s upon recognizing its dangerous effect on human health and ecosystems—allowed global warming to manifest more fully.

"'Global dimming' seems to have gone away in the 1990s," explained atmospheric chemist Meinrat Andreae, referring to a 2007 paper by Michael Mishchenko and colleagues. "People are realizing that aerosols kill. Sulfur, which used to be the main aerosol pollutant, is becoming less and less important because that's the one thing we know how to clean up," Andreae said. Many developed countries have smokestack technology that collects most sulfur compounds before they can become airborne. Even China, a country where urban rivers run black and a brown cloud usually hovers over its capital, has begun to clean up its pollution.

As pollution control kicked in during the 1970s, the dimming that had masked the temperature rise faded away. By the end of the century, the warming since the mid-1970s stood out as unusual. From the outskirts of the Arctic to the peaks of the Andes, normally frozen locales were thawing out. Africa's Mount Kilimanjaro lost about 80 percent of its ice between 1912 and 2000, research by Lonnie Thompson of Ohio State University and others showed. Changes in precipitation patterns and other factors may be adding to the glacier's decline. Still, evidence suggests that modern temperatures are reaching unprecedented highs, Thompson said during a March 2007 talk at the University of Arizona. "What really brings the message home is what is happening to the ice itself," said Thompson, who by then had led or participated in 50 expeditions to polar ice sheets and tropical glaciers. The modern melting even makes the ice look different. When ice melts and refreezes, it contains misshapen air bubbles, he explained. He and others have found these elongated mutant bubbles in the top layers of the Kilimanjaro ice cores. But they show up only in ice formed during the past few decades—nowhere else in ice cores that stretch back eleven millennia, he said. "As we go back in time," he added, "the unusual nature of our current trend really stands out."

Debate continues over how the modern temperature rise compares to temperatures during the Medieval Warm Period. During this time, which arguably centered on the years 900 to 1200, temperatures in Greenland and some parts of Europe were unusually warm, perhaps even comparable to modern temperatures in some locales. Other factors besides temperature make the modern warming different from the historic Medieval Warm Period, though. The ongoing global warming is occurring simultaneously across the planet, while the Medieval Warm Period centered on Europe and the North Atlantic, with spottier warming episodes in other locales. Also, the current warming is linked to an excess of greenhouse gases, while Michael Mann and colleagues suggested in a 2005 *Journal of Climate* article that the Medieval Warm Period related to an increase in solar power coupled with a decrease in volcanic eruptions. That brings us back to the issue of how volcanic emissions and pollution act similarly to shield the planet from the full impact of solar heating.

Pollution acts like a smoke screen in more ways than one. Not only does it shield the planet's surface from some of the ongoing warming, it adds confusion about the extent of warming we face in the future. That is, if aerosols have been masking the past warming more than calculated, the future warming could be worse than we think, as Andreae's 2005 research with colleagues indicates. Any relief obtained from future volcanic eruptions, nature's main source of sulfate pollutants, would be temporary.

Whether the emissions come from volcanoes or air pollution, warming is inevitable at some point. Both volcanoes and fossil-fuel burning serve up a hefty share of greenhouse gases along with air pollutants. And while the sulfate particles from pollution come back down to earth in a week or so, and those from powerful volcanoes typically descend within a year or two, greenhouse gases stay aloft for decades. Both pollution sources and volcanoes spew out enough carbon dioxide and other greenhouse gases with these cooling aerosols that their long-term effect on climate is to warm it.

❧

The same factors affecting our own ability to stay comfortable on a hot day influence the temperature of a planetary body. Our comfort level relates to the sun's intensity, how much of that solar power is collected versus reflected by our bodies, and the presence of water and other greenhouse gases. Clothes affect our ability to reflect the sun's rays. Tennis whites are cooler than that heat-absorbing black number. Trees, roofs, even clouds also act as a shield between the sun and the surface of your skin. Spraying water around also helps, through evaporative cooling. Once the sun goes down, though, having extra water vapor in the air can backfire. Anyone who has tossed and turned through a humid August night recalls the heat-trapping capacity of water vapor. Not only does water vapor contain latent heat that we can sense at some level, it's also the world's dominant greenhouse gas. Thus, the cloudy sky that deflected sunlight like an aluminum mat becomes a nighttime blanket that traps some of the heat that did get through.

The planet responds to these same climatic influences of reflectivity, intensity of sunlight, humidity, and greenhouse gases—on many different time scales. All these factors come into play in the shifts between ice ages and the warmer intervals known as interglacials, as well as in the even warmer hothouses of the past 100 million years. Reflectivity affects a planet's temperature. (Scientists, though, prefer to call this factor *albedo*. The higher an object's albedo, the more sunlight it reflects.) As with clothing, the reflectivity factor often comes into play in black-and-white terms on Earth. White snow reflects more sunlight than dark soil. Skiers know this—they often wear sunglasses in the middle of winter because of the harsh glare reflecting off the slopes. Icy tundra can shoot back three-quarters of the solar radiation striking it. Once the ice and snow melt, though, the tundra's ability to deflect incoming rays from space drops down to about 10 or 20 percent. That's because soil generally looks dark from space. So does vegetation. Spruce and pine forests, like tundra, absorb most sunlight coming their way once they've shaken the snow from their branches, researchers led by Werner Eugster found. Similarly, lakes and seas absorb almost all the sunlight that reaches their surface. If they didn't, a day at the beach might be unbearable on the eyes. As it is, light-colored sand reflects enough sunlight to make sunglasses more than a fashion statement in beachwear.

The lack of reflective ability of open water, soil, and vegetation understandably raises concerns that global warming will speed up as snow cover shrinks. Ice sheets covered about 25 percent of the continents during the coldest part of the last ice age (about 20,000 to 30,000 years ago), even counting the extra land exposed as the ice sheets sipped up the ocean. This compares to 10 percent of land they cover today. The white cloak helped keep Earth frigid during the ice ages, by deflecting sunlight even as it chilled the air. Our modern ice shield is shrinking fast in summer, just when we most need its sun-deflecting skills.

Clouds also reflect some incoming sunlight before it reaches Earth's surface. They currently deflect about half of incoming sunlight in humid regions around the equator, for example. The reflectivity issue also explains why predicting cloud cover remains a Holy Grail of global climate modeling. Modelers struggle to consider how, when, and where cloud cover will change with warming and how much this might moderate temperature rise. Future cloud cover might block some rays from penetrating into the areas below their protective shield as our planet warms. But clouds also radiate heat toward the surface and help trap it there for another round. A cloudy day typically doesn't feel as hot as a sunny day, whereas cloudy nights don't cool off as much. All that water vapor hovering over an area traps heat. So, do clouds have an overall cooling or warming effect? It depends on cloud type, height in the atmosphere, and time of day, among other things. Looking at clouds from both sides, though, we can't assume future protection from warming. Greenhouse gases will be greenhouse gases, after all.

Reflectivity affects the amount and intensity of sunlight reaching Earth's surface. Solar cycles, meanwhile, affect the amount and intensity of sunlight reaching the top of the atmosphere. Solar variations tend to follow relatively predictable cycles, such as those governing our clocks and calendars. We all know that temperature rises and falls daily with the sun. In Tucson, the daily temperature often drops by 30 degrees

Fahrenheit or more after the sun sets. Nighttime temperatures drop less dramatically in humid places or seasons, but at least the solar heating takes a break. The sun's influence defines the seasons as well. As Russian climatologist Mikhail Budyko pointed out in his 1982 book, *The Earth's Climate*, the origin of the word "climate" traces back to the Greek word meaning "to tilt." The planet's tilt defines the seasons by shifting different parts of the planet into the most direct path of incoming rays. As Earth tips its Southern and then Northern Hemisphere toward the sun, the direct line of fire migrates from Rio de Janeiro to Havana. In Chicago, a beam of sunlight at high noon packs more wallop per square inch during the summer solstice than the winter solstice. That's partly why Chicagoans can do a slow burn in summer yet face frostbite in winter.

Besides seasonal cycles, other solar variations make their marks on the planet. Sunspots affect the intensity of sunlight reaching the planet from decade to decade. Stars, and the sunlight they shed on their planets, strengthen gradually over billions of years. Our sun's average intensity has risen from about 325 watts per square meter (roughly a square yard) half a billion years ago to preindustrial values of about 342 watts per square meter. The difference amounts to less than a 20-watt light bulb per square meter, but it adds up across the globe. The sun's growing warmth in time makes it all the more surprising that Earth managed to get colder overall these past few million years.

The recurring ice ages of the past several million years relate to other types of changes in the amount of solar energy reaching Earth: Milankovitch cycles. These cycles take their name from a Yugoslavian astronomer, Milutin Milankovitch, who spent decades of the early twentieth century hunched over a desk scribbling out the complex calculations behind the fluctuations. Even when the Austro-Hungarian Army put him in prison during World War I, he continued to work out the mathematical details about how neighboring planets influence Earth's sun cycles. This is not the Saturn Return and Jupiter Rising of astrological lore. Astronomically, the movements of these giants and other planets exert a subtle and predictable influence on Earth's orbit—and therefore its climate. The cycles change with the shape of Earth's orbit, the tilt of its rotating axis, and whether the land-heavy Northern Hemisphere comes closer to the sun during summer or winter. Along with greenhouse gases, these three cyclical changes combine to create ideal conditions for ice-sheet buildup, then align to promote a full or partial meltdown. Milankovitch focused his calculations on the edge of the Arctic Circle, pinpointing 65 degrees North as key to the forward-backward two-step of northern ice sheets. For the past half-million years, Milankovitch cycles have been clocking in major ice-sheet development about every 100,000 years.

❧

Although Milankovitch wasn't around to see it, his work took a leap forward in 1976. That's when geoscientists James Hays, John Imbrie, and Nicholas Shackleton published a *Science* paper linking physical evidence to his hand-wrought calculations, reconfirmed with modern computers. These three pioneers helped launch the field of paleoclimatology (the study of climate beyond instrumental records). For years, they had proposed that the roller-coastering isotopic signal in deep-sea sediments reflected the comings and goings of ice ages. In their seminal 1976 paper, they supported their premise with strong evidence that the isotopic ups and downs reflected ice-sheet buildup rather than just cold temperatures in the deep blue sea. What's more, their 400,000-year sediment record followed the same time schedule as the Milankovitch variations in summer sunlight at the edge of the Arctic Circle. Their work sent ripples through the field of paleoclimatology.

"Suddenly, Quaternary science was on a theoretical footing," said Owen Davis, a University of Arizona professor who specializes in the climate and ecology of the Quaternary Period, which he described as stretching from about 2.6 million years ago to the end of the last ice age. The theory that Milankovitch solar cycles govern the timing of the ice ages held up through other tests, establishing these cycles as the standard by which almost every long-term record is compared. The work also showed that the four ice ages documented by on-the-ground evidence barely scratched the surface of these ups and downs. The Quaternary featured fifty or more ice ages, depending on how many peaks and valleys are counted in the isotopic record. Apparently the telltale signs of dozens of earlier ice ages, such as piles of rocks from distant mountains, had been buried by subsequent glacial action.

Even as these groundbreaking scientists were setting standards in paleoclimate using sediment cores from the deep sea, other researchers were making new discoveries using ice cores extracted like Popsicles from the ends of the Earth and the highest tropical mountain peaks. These frozen slices of time revealed additional details. Most famously, they revealed the clear connection between levels of the greenhouse gases carbon dioxide and methane and warmer temperatures as glaciers advanced and retreated.

Tropical ice cores carried on the backs of yaks treading from the highest peaks through the hottest jungle helped Lonnie Thompson and Ellen Mosley-Thompson, a longtime couple who together run the Ohio State University's Byrd Polar Research Center, and their colleagues piece together ice-age conditions in the tropics. They and many other international researchers collected cores from ice sheets north and south, traveling in aging helicopters, often sleeping in tents or temporary igloos for months while they drilled cores out of icy expanses near the poles. After carefully transporting these fragile treasures to their labs, researchers would carefully crush sections of the ice to release the ancient bubbles of trapped air, precisely measuring its greenhouse-gas content. In some cases, they also used the isotopic composition of water preserved in the cores to estimate global temperatures for various time spans. By 2007, Jean Jouzel and Jérôme Chappellaz and their research teams had described efforts using Antarctic ice cores to reconstruct temperature and carbon dioxide, respectively, for the past 800,000 years.

All that effort by hundreds of researchers from dozens of countries has turned up surprisingly consistent results: Carbon dioxide levels during the measured time span typically ranged from roughly 170 parts per million during ice ages to 280 to 300 parts per million during warm interglacial periods. The synchronicity between carbon dioxide and temperature adds harsh reality to the scenarios projecting how future temperatures are likely to rise with carbon dioxide levels. As many will recall from Al Gore's compelling demonstration with a hydraulic lift in Davis Guggenheim's documentary *An Inconvenient Truth*, modern levels already are well above typical interglacial levels—and they're still climbing. If we don't change our habits soon, they're headed toward the values of earlier hothouses, when both poles were ice-free.

❧

Carbon dioxide and temperature clearly swing together through the ages. Even so, scientists are still involved in a chicken-egg debate about these two in the ice-core record. You might think it would be easy to see which comes first, a rise in temperature or an increase in carbon dioxide levels. After all, both measurements sometimes come from the same square inch of an ice core, suggesting a match in time as well as place. But an air bubble's content can change with the atmosphere until it's fully sealed in deep ice, a process that European researchers Thomas Blunier and colleagues note can take anywhere from a decade to several millennia, depending on how much snow falls. Snow must cover an air bubble to depths of 150 feet or more before air stops diffusing into its spherical confines. Judging from present-day snowfall, this typically takes about 200 years in central Greenland and 2,000 to 3,000 years around the Antarctic's Vostok station. And it's the Vostok cores that provide the longest records. Ice sheets at both poles measure about 3 kilometers thick (slightly less than 2 miles), but Greenland's higher snowfall rates mean its ice sheet spans a shorter time frame than Antarctica's.

Taking into account the time it takes ice to fully seal atmospheric gases, Blunier and colleagues figure the carbon dioxide rise typically preceded the global warming in the Arctic, and followed it in the Antarctic. The uncertainties about how long it takes to seal the air bubbles make it difficult to draw firm conclusions, though. For instance, a research effort led by Jean-Robert Petit reported that modeled temperature dates on the Vostok core could be off by about 5,000 years. Still, it's generally accepted that the changing solar cycles launched the shift toward a warmer planet, while rising carbon dioxide levels assisted in the meltdown. Warming oceans and permafrost would be inclined to release carbon dioxide as the more intense northern sun heated the land and the air above it.

Carbon dioxide provoked the periodic meltdowns of ice sheets, even if it wasn't the initial impetus behind the temperature rise, Nicholas Shackleton concluded in a 2000 *Science* paper. Shackleton compared data from 400,000 years' worth of ice cores with those from seafloor sediments. Whereas deep-sea sediments generally reflect the amount of water tied up in ice globally, ice-core chemistry tends to reflect local and cloud temperatures. Shackleton's work indicated that the melting of the ice (as registered in the seafloor sediments) lagged behind the rise in local temperatures (as recorded in the ice). Basically, while Milankovitch changes in sun cycles appeared to drive temperature changes, the subsequent buildup of carbon dioxide in the air appeared to trigger the melting of ice sheets. Judging from Shackleton's work and other analyses, it takes time for rising greenhouse-gas levels to fully destabilize ice sheets. But it's not clear exactly how much time and carbon dioxide it would take to bump sea levels back up to interglacial heights of an extra 20 feet.

As with the temperature rise, the sea-level issue comes down to how high and how fast. The sea reached much higher during the Eemian, the most recent interglacial period that stretched from about 130,000 to 118,000 years ago, reminds Jonathan Overpeck, co-director of the University of Arizona's Institute of the Environment and a Nobel laureate as a lead author of the paleoclimate chapter in the IPCC's 2007 report. At an October 2006 global climate change talk in Tucson that drew about a thousand people, Overpeck noted that global sea level ranged from 18 to 21 feet higher during the Eemian compared to the present. "How do we know that? Because there are coral reefs that grew underwater at that time, 125,000 years ago, that are now about that far over sea level. For example, the Florida Keys—you can drive on those now. A hundred and twenty-five thousand years ago, they were several feet underwater. And you can see these kind of reefs all over the ocean at low latitudes, and you can date them very easily."

Many of the coral reef remnants exist far from the influence of earthquakes and other ground-shifting activities that could offer a different interpretation for their exposure. Yet the temperature difference at the Eemian's warmest phase topped today's average by only a few degrees—well within the range expected by the end of this century, maybe even by 2050 given our current emissions of greenhouse gases. Working with colleague Bette Otto-Bliesner and others, Overpeck projected that sea level could rise 3 feet—maybe more—this century. That's higher than the IPCC's 2007 projection for a 2-foot maximum sea-level rise by century's end, although the report acknowledged that this estimate does not include all the factors affecting ice sheets.

Many scientists consider the IPCC estimates for sea-level rise to be on the low side. James Hansen, director of NASA's Goddard Institute and an early proponent of the need to act to forestall global warming, suggested IPCC authors were overly "reticent" in their 2007 report regarding how quickly sea level could rise. In a 2007 article in *New Scientist*, Hansen noted that sea level rose about 65 feet in four centuries during a portion of the meltdown that took the planet out of the last ice age. That averages out to about 16 feet a century, he noted, adding, "There is growing evidence that the global warming already under way could bring a comparable rise in sea level."

Given activity over the past 100 million years, current sea level is in a roughly intermediate stage. At the depths of ice ages, sea level tends to drop by about 300 feet. During hothouse periods, when little if any permanent ice locked up water, sea level rose by about the same over today's levels (although there's a larger spread in the estimated range). The melting of the Greenland ice sheet, the major immediate concern, eventually would raise seas about 20 feet if it continued until only a few scattered shards of ice remained. To bring us back to a hothouse climate would require the thawing out of the Antarctic, something that hasn't happened for a good 35 million years and isn't on the ledger for the immediate future. Portions of it could melt, though, with the West Antarctic ice sheet under the most scrutiny for this possibility. While no one can say for sure what the near future holds, the past makes it clear that ice sheets are not a required feature of a stable climate on Earth.

For long stretches during the past 100 million years, the planet's temperature climbed so high that even Milankovitch cycles of solar intensity caused little, if any, ice to survive the summer heat. This occurred even though the sun as a whole was emitting less heat. Without ice and snow to shield the surface from the summer sun's rays, reflectivity dropped in importance as a moderating factor. The position of the continents was a relevant factor during long-ago glaciations, such as when Australia migrated to the South Pole about 300 million years ago during the Permian. When a major continent drifts toward the poles, it's easier for ice to build up. But by 100 million years ago, the continents had shifted into roughly their current locations, albeit with a thinner Atlantic Ocean at the beginning of that time frame. Given that, greenhouse gases represent the most viable explanation for the exceptional warmth of hothouses during the past 100 million years, such as the middle Cretaceous and early Eocene.

For much of the dinosaurs' reign, an ice-free world with high carbon dioxide levels featured toasty tropical seas and balmy and sometimes scorching conditions over much of the globe. During the Cretaceous, some dinosaurs even roamed the Arctic Circle. "Cold-blooded" (ectothermic) creatures, their veins would have turned to ice in today's polar regions. But the poles looked different during the Cretaceous hothouse, which stretched from about 144 to 65 million years ago. For one thing, massive forests grew around them. There were no permanent ice sheets at sea level, although research by Kenneth Miller and others suggests that midcontinental ice in Antarctica sporadically built up during hothouse "cold snaps." Without ice permanently tying up some of the ocean's water, sea level reached an estimated 150 to 300 feet higher than today, depending on locale and analytical approach, these researchers note.

From a peak in the mid-Cretaceous about 100 million years ago, temperatures gradually declined, with erratic ups and downs on top of the general cooling trend. A secondary peak in warm temperatures apparently arrived with the Eocene about 55 million years ago. Crocodiles, warm-water animals that now rarely venture past Florida and the coastal plains of other southern states, made their way as far north as Canada's Ellesmere Island, Paul Markwick noted in a paper linking the expansion of these reptiles to warmer climates. Their presence in Wyoming waters as well suggests that Cheyenne had at least twice as many freeze-free days than its current balance of about 120 a year. Palm trees and other warm-weather plants from Wyoming's Green River Formation indicate that temperatures averaged in the mid-40s, well above freezing, even during its coldest month. Even the poles were ice-free. Again, forests grew in the Arctic Circle. Apparently, no sea-level ice survived the winters.

Some of this polar warmth might have related to an unexpected source—hothouse hurricanes. Kerry Emanuel, a Massachusetts Institute of Technology physicist who has been documenting the strong influence of sea-surface temperatures on modern hurricanes, also has been modeling how hurricanes disperse heat from the tropics. His modeling work suggests that not only do hurricanes cool the surface of tropical seas, they also move heat downward into the sea. These pockets of hurricane-transferred heat then move by "Rossby waves" to warm up deep waters around the world, he said, likely surfacing at higher latitudes closer to the poles. Interestingly, other researchers have found that the greatest warming of the Atlantic Ocean since about 2000 occurred about midway between the equator and the North Pole, between about 40 and 50 degrees latitude. Using the year 1996 as a case study, Emanuel calculated in 2001 that hurricanes could make a serious dent in the amount of heat scientists know is exported from the tropics. What's more, he has proposed that intense hurricanes could have helped keep the tropics cooler and the poles warmer during the Eocene and Cretaceous hothouses.

Something must have operated to move heat from the tropics to the poles during these hothouse climates. Everyday winds were an unlikely candidate for the role, given that winds thrive on temperature differences—such as the extremes between the equator and poles that occurred during ice ages. Evidence suggests that raging ice-age winds settled down as the temperature at the poles climbed closer to that of the tropics. With this and other factors in mind, Emanuel developed a simple model, published in 2001, that suggested strong hurricanes during hothouse climates could have played a key role in moving heat poleward from the tropics.

Matthew Huber of Purdue University and colleagues have been working on the issue of hurricane heat distribution, too, using computer models to consider the influence of these ephemeral storms on ocean temperatures. Although his recent work with Ryan Sriver focuses on how modern hurricanes cool the tropics, his previous work with Lisa Sloan showed that models of Eocene climate needed mechanisms to warm the poles and cool the tropics. And the Eocene tropical sea-surface temperatures of 86 to 90-plus degrees Fahrenheit, as estimated by researchers Paul Pearson and colleagues, reached high enough to pump up hurricanes.

"Their sea surface temperatures imply more intense storms, and we have found that it is the intense storms that do most of the heavy lifting in terms of raising cool waters in the tropics, thus causing vertical mixing," Huber explained in an e-mail exchange. "Increased vertical mixing can absolutely resolve the long-standing problems in Eocene paleoclimate in conjunction with much higher than modern levels of carbon dioxide (or other greenhouse gases)." Huber suspects that the Eocene's higher levels of greenhouse gases, not hurricanes, had the primary role in keeping the poles so toasty.

❦

The lack of preserved ice beyond the last million years means there are no air bubbles to mine for direct evidence of greenhouse-gas levels during earlier hothouse periods. Researchers briefly had hoped that trapped air in ancient amber could mimic the ice cores by sharing information on ancient carbon dioxide levels from 50 million years ago or more, as Thure Cerling noted in 1989. The golden-yellow gemstone made from tree sap preserves insects so well that fiction writers fantasize about extracting DNA from bugs entombed in amber, as in the Steven Spielberg film that made "Jurassic" a household word. But amber doesn't do carbon dioxide, as researchers were disappointed to discover. The bubbles are just too close to the surface to exclude interference from modern air. So information on atmospheric content beyond the Quaternary Period of the past couple million years must be inferred from less direct means, such as the chemistry of soil nodules, the chemistry of certain algae and liverworts, and the internal structure of plants (chapter 5 will cover the latter). Still, this indirect evidence indicates that greenhouse gases ran high. During the Cretaceous, volcanoes were particularly active, filling the atmosphere with greenhouse gases—mainly water vapor and carbon dioxide. And escaping methane gas may well have launched the temperature extreme of the early Eocene.

A growing body of research into conditions before the recent ice-age/interglacial seesaw of the Quaternary upholds the general relationship between carbon dioxide and temperature. Researcher Dana Royer pulled together close to five hundred carbon dioxide estimates predating the Quaternary for comparison in a detailed analysis published in 2006. As he noted, most of the data points for this half-billion-year stretch had been reported since 1998. Comparing results at million-year increments or more, he found that, where evidence existed, past periods with evidence of cool or icy conditions matched periods of relatively low carbon dioxide levels. Similarly, hotter periods registered high carbon dioxide levels.

Lengthy stretches of extensive ice cover, including modern times, featured levels of carbon dioxide consistently below 500 parts per million. Northern ice sheets become extinct once carbon dioxide levels reach somewhere between 500 and 1,200 parts per million, computer modeling work in 2003 by Robert DeConto and David Pollard indicates. That suggests a likely minimum value for carbon dioxide levels during at least some parts of the Eocene and Cretaceous. It's even within the range of projections modern civilization could reach within a century at the pace projected for an industrialized world. Chalk up another reason it's important to consider how the planet fared before the relatively icy times of the past 3 million years.

The estimates Royer compiled from data generally agree well with a computer-modeling approach developed by Robert Berner. What's more, when Royer, Berner, and another colleague used their independent approaches to estimate the global temperature rise if modern carbon dioxide levels reach 550 parts per million, their 2007 results consistently hovered right around 3 degrees Celsius (about 5 degrees Fahrenheit). "It was right in the middle of what everybody is getting," Berner told me in late 2007, referring to the Intergovernmental Panel on Climate Change's 2007 "best-estimate" climate-model projections for a doubling of carbon dioxide levels over preindustrial values. So, once again, the evidence converges upon a serious temperature rise if we continue to burn oil, gas, and coal. At any rate, there are only so many ways to warm a world.

After solar cycles, reflectivity, and the position of the continents are taken into account, greenhouse gases generally make up the balance.

❧

While scientists are working mainly with indirect evidence when it comes to greenhouse gases during earlier hothouses, they recently found what looks like a smoking gun to explain warmth in the early Eocene. A startling shift in carbon chemistry suggests a rapid burst of greenhouse gases into Earth's oceans and atmosphere right before the Eocene started 55 million years ago. Isotopic records show a sudden peak in "light" carbon, the kind associated with oil and gas. Fossil fuels, including methane, are so old that most of the heavier oddball carbon isotopes have long since decayed, leaving only light carbon behind. The jump in light carbon just before the Eocene goes well beyond what volcanic activity could explain. Scientists are converging upon an explanation involving the massive release of underwater methane—known as natural gas for more reason than one, perhaps.

Could the ancient oceans have burped out methane in a series of eruptions that spurred on the documented warming? There's some ocean-based evidence for this—acidification that led to the mass extinction of about a third to half of bottom-dwelling marine species in existence at the time. Their passing helps mark the boundary between the Paleocene and Eocene. Sadly, the calcium carbonate shells of these creatures may have served as the ocean's version of Tums. After all, waterbound methane can transform into carbonic acid. Even so, much of the methane must have escaped to the atmosphere, as it's not readily dissolved in water. A meteoric rise in atmospheric methane would handily explain the jump in temperatures that occurred soon after the mass extinction of ocean bottom-dwellers. Deep-sea temperatures globally rose by 9 to 11 degrees Fahrenheit in roughly 10,000 years, a blink in geological time, as James Zachos and others describe in a 2001 *Science* paper. The Intergovernmental Panel on Climate Change considers methane to be about twenty-five times more powerful in warming the atmosphere than carbon dioxide, when comparing one molecule to another. Airborne methane lasts only about a decade or two under current conditions, but then it transforms into carbon dioxide and water, both effective greenhouse gases in their own right.

Interestingly, other researchers found a similar "light" carbon spike about 117 million years ago, during the Cretaceous. Evidence compiled and published in 2001 by Hope Jahren of John Hopkins University and Nan Crystal Arens of the University of California–Berkeley and others suggests that there was a rapid release of the kind of carbon found in fossil fuels. It would take a tenfold increase in atmospheric carbon dioxide to explain the sudden change in the carbon isotope signal they picked up by analyzing ancient plant parts. This is almost certainly beyond what volcanoes can do, they noted, even given the abundant volcanic activity during this time frame. It's well within methane's reach, however. A release of only about 2 percent of the methane pockets stored in today's oceans would explain this Cretaceous peak in "light" carbon, they noted. Along with the inferred long-term increase in greenhouse-gas levels from the many active volcanoes, this apparent outburst of methane would help explain the exceptionally warm times of the Cretaceous, when evidence suggests the globe's temperature averaged up to 27 degrees Fahrenheit higher than today's average.

Temperatures soared above today's values even before these apparent methane bursts, though. We can take some comfort in that, perhaps. Researchers suggest that the extra heat might have helped destabilize these underwater pockets of natural gas. There's an eerie similarity, though, between our rate of fossil-fuel release and the apparent release of methane hydrates just before the Eocene, as Ellen Thomas and her colleagues point out in a 2000 book, *Warm Climates in Earth History*. Already, modern carbon dioxide levels are nearly 390 parts per million, about a third higher than the 270 or so parts per million measured in air bubbles from interglacial warm periods, including our modern one before the industrial age began. Levels had already climbed to about 315 parts per million by the time Charles Keeling and his colleagues began measuring carbon dioxide with instruments in 1958. Still, the methane and carbon fossil fuels powering the ongoing industrial age could become a short-lived, almost ephemeral, spike on geological scales if society decides to switch to renewable sources instead of fighting for fossil fuels.

As it is, our fossil-fuels habit is pushing our greenhouse-gas levels higher than they were during the last interglacial. On a positive note, our current cycle of solar intensity resembles the configuration when the last interglacial was ending, as researchers George Kukla and colleagues note. This means the sun's intensity is on a waning cycle, which theoretically could help balance higher carbon dioxide levels on geological time scales of thousands of years. In human time frames, though, we can expect to face many challenges from rising greenhouse-gas levels and the higher temperatures they bring, along with physical responses such as melting glaciers, rising seas, and more intense storms. Like hurricanes, these physical responses also can play a Gaian role in moderating the temperature rise.

❦ 4 ❦

Circulation Patterns

A Pulse of Floods, to the Beat of Rising Waves

By mid-June, I often wonder how I ended up living in the desert. Tucson temperatures regularly soar to 105 degrees or more, making daytime bicycling an exercise in endurance. In 2007, we faced thirty-nine consecutive afternoons topping 100 degrees before temperatures briefly dropped below the century mark. In any June, the air weighs heavy with dust. The sun is omnipresent, its rays penetrating cloud-free skies to zap any skin left exposed. Finally, as if Nature can't stand it any more either, the monsoon comes to the rescue. The winds shift. Now they're coming from the south, suffused with moisture from distant seas. Clouds gather in afternoon councils, conferring in rumbling voices about what to do next. In the interim, they scatter and deflect the sun's rays like an opaque umbrella over the desiccated landscape below. Finally, the first raindrops fall in lukewarm splatters on skin, concrete, and prickly pear. The moisture releases the aroma of creosote, the signature scent of the desert springing back to life. *Eau de vie.* Within days of the first good rains, grasses and wildflowers lend color to the toasted landscape. The annual floods begin.

Monsoons, like hurricanes, exemplify how warm temperatures can inspire more rainfall and flooding, even if for a season. The thunderclouds announcing monsoon season typically arrive around the summer solstice, much like their brethren that gang up into tropical cyclones. As with hurricanes, rising sea-surface temperatures may even spur more intense monsoonal rainfall in some cases. The heating of the land even more clearly helps to pull in the winds and rains that mark the monsoon. Scientists have recognized this dynamic since at least the late nineteenth century, when Henry Blanford observed that India's annual monsoon tended to weaken when years of abundant snowfall on the nearby Tibetan Plateau moderated the seasonal warming.

In fact, warmer temperatures bring on more rain and snow in general, with precipitation more likely to fall with greater intensity. For one thing, warm air holds more moisture. Other factors also contribute. The connection between warm climates and more intense precipitation operates across time as well as space, as this chapter will illustrate. More rainfall tends to mean more floods. Keep in mind, though, that "more rainfall" globally doesn't translate to "less drought" regionally. Global warming and related climate shifts can exacerbate drought in some regions, possibly including the U.S. West. Ironically, even drought can encourage damaging floods, such as by killing plants that otherwise would have tempered the impacts of the next big rain. Clearly, rising seas will exacerbate the potential for more floods as the planet warms. Warming seas that promote stronger hurricanes bring a greater risk of floods—not only along the coast, but anywhere remnant hurricanes carry moisture inland. So does warming air, because of its ability to lift and shift more moisture.

When monsoon moisture hits the ground, dry riverbeds can spring to life for hours or days. City streets overflow with the bounty. From a rented house in Tucson's downtown area, I watched plastic bins and other flotsam and jetsam drift down Mabel Street, transformed a dozen times a summer into the River Mabel with knee-deep flowing floodwaters. These monsoon cloudbursts account for half the city's rainfall in a typical year. Granted, that's still only about 6 inches of rain. The more famous Asian monsoon ushers in about three-fourths of India's yearly rainfall—often in dramatic downpours, such as the 40 inches that fell on Cherrapunji on one June day in 1876. And that's only about one-tenth of the yearly rainfall Cherrapunji averages, mostly during the monsoon season. Even Tucson's monsoon storms can be damaging, as in the July 31, 2006, flood and debris flow that eroded roads and changed the configuration of Sabino Creek, an oasis in the Santa Catalina Mountains to the city's north.

Around the world, floods affected more than 1.5 billion people in the quarter century ending in 1995. They killed more than 318,000 people and left 81 million homeless. Globally, there's a strong link between temperature and "runoff"—water flowing toward streams, perchance to flood. Wet conditions almost always came with warmer-than-average years for the world as a whole, based on an analysis by Jean-Luc Probst and Yves Tardy that covered the beginning of the last century through the mid-1980s. Similarly, dry conditions came with cooler-than-usual years for the planet as a whole. (Regionally, and locally, some droughts and floods ran counter to the global pattern, of course.) Researchers have detected a trend toward wetter times as the U.S. climate warmed as well. Over the twentieth century, precipitation increased by about 7 percent in the continental United States with the eastern half registering the biggest increase, an analysis by Pavel Groisman and colleages found. They also found that the most extreme storms (those in the top 1 percent of the record) increased by a full 20 percent.

This increasing precipitation is empowering floods. U.S. flood damages increased by an average of 3 percent a year between 1932 and 1997, based on an analysis by Roger Pielke Jr. and Mary Downton that adjusted for inflation and a growing population. And that was before Katrina's flooding broke cost records for damage. Specifically, Pielke and Downton found that flood damage across the nation increased with the number of rainy and snowy days, especially when multiple unusually wet days occurred in a row. (The "unusual" aspect of the interpretation related to norms by region.) Their conclusion supports what most people would guess—flood damages increase with precipitation. Because this connection holds logically as well as analytically, most of this chapter will focus on why precipitation tends to increase as the world warms. It goes mostly without saying that increased precipitation translates into increased flood risk as temperature rises, but that's an important underlying reality. What's more, there are Gaian values to increased rainfall and flooding.

❧

The type of land heating that promotes summer monsoons can kick up thermal winds on any hot day in semi-arid lands. My colleague Elise Pendall, who now runs an isotope lab at the University of Wyoming, drew this to my attention one April day while we were both graduate students collecting soil and plant data in experimental wheat fields south of Phoenix. As we worked through the rows, the sun would hammer us relentlessly all morning. Regardless of what the calendar says, you know it isn't spring when the temperature approaches 100 degrees Fahrenheit. Breathing the air was like inhaling dead weight. Finally, just before noon, local winds would rush in to liven things up. A long morning of baking the land had stirred the deathly still air parcels into an uprising. As they lifted to greater heights, nearby air would whoosh in to fill the void. These thermal winds brought welcome relief from the oppressive heat.

A similar mechanism at the regional scale launches the summer monsoon, which arises when winds move to fill the void created by hot air rising over superheated land. The term "monsoon" refers specifically to a change in wind direction. (In this book, "monsoon" will refer only to summer monsoons, which carry sea-moistened air toward land.) Water warms more slowly than land, in part because there's no shortage of fodder for heat-trapping evaporation. Thus a ray of sunshine vaporizing water retains less energy to heat the surrounding air, land, and sea. But the warm, cloudless spring days typical of monsoon-prone lands leave behind little moisture to temper solar power. In the subtropical regions prone to monsoon activity, pre-monsoon days regularly climb above the 100-degree-Fahrenheit mark. (Note that this book takes a view of "subtropical" conforming to geography rather than ecology, defining the subtropics to be the region on the globe between roughly the edge of the tropics to about 40 degrees in latitude.) The general lack of clouds and moisture means most of this energy goes into toasting the land. Once the land heats up to a certain point, it begins to vacuum in maritime air. The blessed monsoon begins.

Monsoons grace subtropical South America, Australia, and Africa as well as Asia and North America. Anywhere that land regularly heats up faster than nearby seas, the shifting winds that define the monsoon can arise. Sometimes, mountains act as accomplices in the wind shift. Mountainsides slant into sunlight. With the proper angle, mountainsides can concentrate solar power into the type of direct beam not otherwise found outside of the tropics. Thus, they heat up faster than the flatlands below that are collecting oblique sunlight. Air swoops in to fill the void created by air rising over superheated mountainsides. From there, it's the standard story: Rising clouds cool with height, sometimes to the point that they drop some of their moisture as rain or snow.

Mountains promote precipitation in another way, too. They present a formidable front that can fling air parcels high into the sky. If the air parcels reach heights cool enough to condense their inner water vapor, they can turn into storm clouds. Mount Waialeale, located on the tropical island of Hawaii's Kauai, averages more than 400 inches of rainfall a year. That's about fifty times the annual rainfall in Phoenix. But even in the Southwest, much of the precipitation occurs around mountains. That's one big reason southwestern forests exist mainly on mountaintops.

A visual display of the power of mountains in making rain surrounded me one spring day while I was walking my dog in a park in central Tucson. While Pepita was off scrounging for pork chop remnants, I looked around and realized I was standing in the middle of a textbook example of the so-called orographic effect. Over the Tucson Mountains to the west, slate gray clouds obscured the yellow-orange setting of the sun. To the north, cumulus clouds crept over the Santa Catalina Mountains. An apartment building blocked my view of the Rincon Mountains to the east, but a gathering of clouds marked their presence like smoke signals in the sky. Looking up, I noticed that a semicircle of clouds had drifted from the Catalinas to lurk over my head. I called my dog over and headed home, thankful that we have mountains to break up the monotony of the landscape and the dryness of the climate.

The mountain habit of promoting precipitation has a couple of downsides, though. For one, the slope downward of prevailing winds often goes lacking. Once the mountain peaks have shaken down the clouds, the impoverished parcels of air start their descent. Descending air typically warms up, so the former cloud now might be stealing moisture from its environs. The "rain shadow" of a mountain's downwind slope often features barren land or desert scrub. For example, the Cascade Mountains collect 60 to 100 or so inches of Pacific precipitation a year along their peaks, leaving less than 10 inches a year for the downwind sagebrush in eastern lowlands of Oregon and Washington. Another disadvantage is that mountains can be overly successful at promoting rainfall, prodding clouds to release the kind of downpours that cause floods. They also fit the profile for landslides, which typically require slopes steeper than 30 degrees. Nicaragua's mountains punched up rainfall tallies during Hurricane Mitch. The resulting landslides brought tragedy to thousands of people living on its steep slopes. In another event, Jamaica's Blue Mountains coaxed 97 inches of rainfall out of a 1909 hurricane. Southwestern mountains have less moisture to collect, but they wring out what they can when it passes by—sometimes in the form of remnant hurricanes. Mountains everywhere provoke rainfall from hurricanes, monsoons, and other convective storms.

❧

Those who ignore folk wisdom to watch a pot of water boil can see how convection works. On the stove, roiling bubbles carry energy from the hot surface to the air above. In the atmosphere, heated air rises in discrete parcels, like giant hot air balloons without the cloth wrapper. These air parcels rise like bubbles as long as they remain warmer—and thus lighter—than the surrounding air. With the addition of water evaporated from nearby seas, the surface heating can lead to the so-called tropical convection that creates thunderstorms of every type, including those characterizing the monsoon. Water vapor inside the air parcels helps lift clouds up, up, and away. When the vapor drops back into water, the energy it releases turns back into heat. This helps power the parcels' continuing ascent. As these buoyant parcels of energized air rise, they face increasingly chillier conditions. The atmosphere cools with height in the troposphere, where weather reigns. At some point, the air will be too cold to hold all the moisture it collected near the warmer surface. Even more of the vapor transforms back into liquid. Down it pours.

Monsoon rains, and thunderstorms in general, often come in the afternoon. In San Juan, we joked about setting our watches by the 3 P.M. storm clouds that would gather over the eastern side of Puerto Rico. The timing coincides with the heating of both the land and ocean surface. The hottest part of a typical day falls between about 2 P.M. and 4 P.M., not noon. In the daily cycle, the sun may be most intense at noon, but the surface heating reaches its peak after several hours of baking under the midday intensity. The Luquillo Mountains no doubt helped maintain this predictable pattern.

Sea-surface temperatures often follow a similar daily cycle, something that sunk in during a warm March 2006 visit to the Gulf of California. After an eight-hour drive from Tucson, I jumped in the water soon after we arrived at a Mexican beach along El Golfo de Santa Clara. At 2 P.M., the glassy green water enveloped me in warmth. After enjoying a lengthy soak, I eventually retreated to land. It was 6 P.M. before I returned to the sea. Remembering my earlier bathing experience, I had expected warm water to make up for cooling air. But its temperature had dropped with the setting sun. The windy day probably sped the cooling, too. The water felt refreshing, but didn't invite lingering. By 9 A.M. the next morning, the sea had cooled even further, enough that I didn't warm up during my ten-minute sojourn into the waves. Experiencing these sea-surface temperature variations firsthand helped me understand why warm-weather thunderstorms often come in midafternoon.

Warming seas can boost monsoon rainfall at the seasonal scale, too. Atmospheric scientist David Mitchell and colleagues used satellite imagery to estimate sea-surface temperature for several years in the Gulf of California, a slice of sea between mainland Mexico and Baja California small enough to fall below the radar of global climate models. They found that sea-surface temperatures there linked to monsoon dynamics in the U.S. Southwest. Sonoran Desert rains would kick into gear soon after the Gulf heated up to about 80 degrees Fahrenheit. During a telephone conversation with Mitchell, I mentioned the similarity between his results and other research showing that it takes a temperature of about 80 degrees to give rise to hurricanes. "It's a very interesting number," he agreed. "That's when we get convection started along the coast of Mexico, and all over the world, really." He referred me to an analysis by Chidong Zhang that found tropical convection, in general, blossomed once sea-surface temperatures reached that 80-degree threshold. What's more, the Mitchell team found that the heaviest daily rainfall tallies in Arizona came once the northern Gulf's sea-surface temperatures topped about 83 degrees Fahrenheit. If that number also sounds familiar, it's because that's about the temperature threshold associated with major Atlantic hurricanes. Clearly, warm seas and intense storms go together at various scales.

❧

While warming land and sea can promote convective storms, warming air can increase all types of precipitation. That's because warm air holds more moisture than cool air. In Tucson, this truth comes home every winter when I must slather on hand lotion several times a day to keep my skin from cracking in the dryness. Meanwhile, it might take all day for my clothes to hang-dry. In July, a shirt hanging on the line often dries within the hour, sometimes within fifteen minutes, but my hands don't need moisturizer. Both dry winter air and high summer evaporation rates relate to how the atmosphere handles moisture. Its water-holding capacity roughly doubles with each temperature increase of about 20 degrees Fahrenheit. The actual rate varies, with greater moisture increases per degree of change as the air warms. That's why global warming is expected to cause bigger increases in evaporation rates and air moisture in the tropics and subtropics than in northern climes, even though the latter is warming relatively faster.

High temperatures boost intense rainfall, whether local or afar. When raindrops fall in warm climes, they can gain size and strength as they encounter vast reserves of airborne moisture on their way down. This applies to warm-season rainfall in temperate as well as tropical and subtropical regions. In an analysis of data from one hundred weather stations around the world, Thomas Karl and Kevin Trenberth found that intense rainfall events of 2 inches or more in a day were several times more likely during warm seasons than relatively cool seasons. In their analysis, average temperatures ranged from 84 to 95 degrees Fahrenheit during warm seasons and 27 to 66 degrees during cool seasons.

The rule that warmer air holds more moisture goes back to basic physics. There's no controversy among scientists over this. There's plenty of debate, though, over where and when that airborne water will come back to earth.

"There are two issues. One is how much moisture is in the air, and the other is what is the mechanism to get it out of the air," Kelly Redmond, a climatologist with the Desert Research Institute in Reno, Nevada, whose cheerful demeanor and detailed knowledge make him a favorite speaker at climate gatherings, told me during a telephone conversation. The Persian Gulf is often humid, especially in summer, yet it rarely rains. "There's no rising motion, no way of cooling the air to get the moisture out," he said. In the U.S. Southwest and other regions favored by monsoons, mountains often provide the uplift mechanism. Wedges of cold air, such as an Arctic front from Canada or Alaska, can act like mountains, serving as a physical barrier to lift warm air aloft. "A cold air mass will help lift the warm air it's coming into, to lift it up and cause cooling and condensation," Redmond explained. The rising warm air then has a better chance of reaching heights cool enough to promote precipitation. Cooling nighttime temperatures can also encourage rainfall, as the loaded atmosphere drops below temperatures that can keep the moisture in its vaporous form.

Even the frontal systems that bring snow and ice generally get their moisture from warmer climes, Redmond noted. A 1999 analysis by Kevin Trenberth, a climatologist with the National Center for Atmospheric Research, suggests that only about a fifth of local precipitation comes from water evaporated within roughly 600 miles of the storm. The rest comes from distant seas and lands, mainly the tropics and subtropics. U.S. rain and snow tend to come from the subtropics (geographically, from the edge of the tropics at 23.5 degrees to about 40 degrees North and South). Within a day or so, moisture from the Gulf of Mexico and subtropical Atlantic can reach the Ohio River Valley, based on Trenberth's analysis, while subtropical Pacific moisture can reach the Rocky Mountains. So warmer subtropical seas can translate into more snowfall over Cleveland or Denver when conditions are right.

As the climate warms, larger sections of the Arctic Sea thaw out earlier in spring and stay ice-free longer in fall. Lakes also stay ice-free for more of the year. In some cases, this, too, can translate into more snowfall in some areas because an unfrozen surface exposes water to evaporation. A 2003 study led by Adam Burnett of New York's Colgate University found a trend toward increasing snowfall since 1951 for some eastern U.S. weather stations that register a "lake effect." The trend only showed up in stations downwind of the Great Lakes where the ice-free source of abundant moisture counteracted the shortening snow season.

In the West, however, many landlocked weather stations are registering a downward trend since 1950 in the amount of snow capping mountains, as Phillip Mote and colleagues have documented. A greater share of precipitation is falling as rain rather than snow, a study led by Noah Knowles found. The snow that does fall is melting earlier. Western stream gauge records show river heights reaching their peaks earlier in the season, with March flows increasing as June flows decline, Iris Stewart and colleagues showed. These changes in snow cover have been linked to an increase in big wildfires in western forests, as mentioned in the introduction. The changes also increase the risk of drought, as snow serves as natural water reservoirs for this semi-arid region. At the same time, the risk of floods rises in the western as well as the eastern United States with global warming. The odds of an intense storm event increase in a warmer climate, as does the potential for heavy rains to fall on melting snow.

Some of the increased risk of floods goes back to basic physics, which applies to snow as well as rain. The moisture in snow generally rises with the storm's overall temperature. A standard rule suggests that it takes 10 inches of snow to yield the liquid water equivalent to 1 inch of rainfall. More detailed measurements show that it actually takes about 13 inches of snow, on average, to make 1 inch of rainfall equivalent, Redmond said, although this varies widely—often with temperature. For instance, it might take 40 inches of "cold snow" to equal an inch of rainfall. "The really powdery white snow that skiers love, up in the high altitudes, that's cold snow," Redmond said. Powder is not the norm, though. "Snow usually occurs mostly between temperatures of 20 degrees and freezing, because the air can hold more moisture," he added. "I haven't seen a really heavy snow when it's below minus 20 or 25 degrees Fahrenheit."

Chicago's worst snowstorm is a case in point. Soon after my conversation with Kelly Redmond, I received an e-mail message from my dad commemorating the fortieth anniversary of this memorable example of warm-weather snow. Jim Allsopp, a meteorologist in Chicago, had compiled old pictures and a description of the storm for the National Weather Service's Web site. The 1967 blizzard dropped 23 inches of snow between early morning one January day and late morning the next. High winds multiplied the effect, piling drifts of snow taller than an adult in some places. Two days before the blizzard arrived, that January 24 set a record for high temperatures that stands today. Witnesses reported funnel clouds, and thunderstorms spotted the city during the evening of that unseasonably warm 65-degree day. Even on the day of the storm, Meigs Field featured thunderclouds. The lake effect added more impact to the main event, a clash in which a cold front from Canada pushed underneath and elevated warm air from the Gulf states. At the time, I was a kid living in Calumet Park on Chicago's South Side. My dad and sisters and I forged a maze of passageways through the snow in our front yard. This snow worked for everything—forts, ammunition, snow-people—it was so heavy and sticky. From my childhood perspective, it was loads of fun. The people in the twenty thousand cars and eleven hundred buses stranded alongside the useless roads probably remember it less fondly. In fact, sixty people died from the challenges, which included a lack of heating oil as delivery trucks became snowbound. Businesses lost the equivalent of $900 million, based on 2006 values. So, once again, a precipitation boon does have drawbacks from a human perspective, even if it makes sense from a Gaian perspective.

More moisture in the air means more water comes back down to earth, whether as rain or snow. So it's clear that warming temperatures will convert to more precipitation—at the global scale. At the regional scale, results will vary. For instance, the 7 percent increase in U.S. precipitation over the last century favored the eastern half of the continent, neglecting the Southwest altogether, according to a detailed 2004 analysis by Pavel Groisman and others at the National Climate Data Center. Hot temperatures globally still can translate into low rainfall regionally—especially in the subtropics. Arizona's Sonoran Desert is a case in point. Outside of the annual monsoon season or an El Niño winter that lives up to expectations by bringing more precipitation, the desert can go without measurable rainfall for months on end. As it happens, even the presence of deserts in the Southwest—and the projection for more droughts there and in other subtropical regions—has its roots in warm surface waters and intense sunlight at the tropics, via the dynamic known as Hadley circulation.

❧

Flying north from Tucson on a dry December day, I could see the results of Hadley circulation spread across the landscape below. Dust-colored hills pocked brown, barren plains. The sandy soil furrowed into formations reminiscent of cauliflower from this altitude. The empty channels below formed the same pattern of branching lines so ubiquitous throughout nature, connecting twigs to trunks, capillaries to arteries, creeks to rivers. In a wetter environment, streams would flow through these cracks in the landscape. In the desert below, no water broke through to the surface. But some lines contained splotches of green, shrubs proving that water does gravitate toward these gullies and depressions during occasional rains. Most of the time, though, Hadley circulation keeps the pressure on and the skies clear.

The desert conditions that stand out when flying at the height of upper-level winds trace back to the contrast between the steamy equator and the freezing poles. Currents of air and water flow up from the tropics in a constant effort to bring balance to these two extremes, driving global circulation in the process. The reigning theory suggests that Hadley circulation evolved because the tropics receive a greater share of solar energy than other parts of the globe. Contrary to popular belief, the sun is never directly overhead in Los Angeles, Chicago, or Miami. Not even at noon. Because of how Earth slants around the sun, sunlight will fall directly only within the tropics, located between the Tropics of Cancer and Capricorn (both at 23.5 degrees latitude). When the sun reaches its northern apex on the summer solstice, a direct ray of sunlight may impart as much energy to one square foot of ground in Havana as a slanted ray might shed on three or four square feet in Anchorage. (The energy difference gets more complicated when comparing the lengthy summer days of high-latitude Alaska to the relatively short summer days of tropical Cuba, but that's a different topic.) All that excess energy shining on the tropics throughout the year has to go somewhere. It heats up the land, especially where moisture is sparse. It warms the ocean, energizing currents. It generates storms as water droplets evaporate, congregate, and head out, often bound for cooler regions. And it heats the air, empowering it to hold more moisture—and also to rise to greater heights. The atmospheric layer containing weather, known as the troposphere, reaches twice as high around the equator as it does at the poles. Thus, tropical clouds are more likely to attain heights that spur rainfall.

Where bands of clouds converge around the equator, the areas below regularly get rain two hundred or more days a year. This roving band of equatorial clouds, known as the tropical convergence zone, marks the tallest peak of the atmosphere's weather layer. The zone migrates around the equator, ranging from near dead-center during the Southern Hemisphere's summer to about 6 degrees North when the Northern Hemisphere faces summer's intense heating. The zone favors the north because of its abundant land, which heats up faster and hotter than the ocean. In fact, the tropical convergence zone can travel far from the equator while passing over land, encompassing Africa's Congo region, India, and much of the Amazon. Over the ocean, the zone seems to migrate with maximum sea-surface temperature, judging from a modeling study that supported observations to that effect, as a *Journal of the Atmospheric Sciences* paper by B. N. Goswami and colleagues reported. So once again, a warm sea surface stimulates clouds and rainfall.

What goes up must come down. Eventually the towering air parcels rising vertically above the tropics lose steam and begin to descend—over the subtropics. Cooler now and often spent of their moisture, they usually heat up and dry out further on the way down, as descending air does. This descending branch of air weighs heavy on the land below. This high-pressure branch of Hadley circulation helps explain the seemingly endless supply of sunny days found in the U.S. Southwest, southern Australia, parts of South America, and, of course, North Africa. These subtropical lands, centered roughly at 30 degrees North and South, feature many of the world's deserts: the Sonoran, Mohave, Atacama, Kalahari, and Sahara. In the Sonoran Desert, the saguaro cactuses that symbolize the U.S. Southwest owe their presence largely to Hadley circulation. With all those sunny days and sparse rainfall keeping many other plants at bay, cactuses thrive in the desert flatlands.

A similar circulation pattern helps explain why the "mid-latitudes" tend to support wetlands. After sinking in the subtropics, air rises again in the mid-latitudes—the region from about 40 to 60 degrees—creating low-pressure zones that promote precipitation. The northern mid-latitudes encompass many humid U.S. cities, in and around a cross-country line from Portland to Boston. Air sinks again around the poles, which look deceptively less arid than they are because snow piles up over the years. The cooler conditions help keep soils moist around the poles and mid-latitudes. Still, the months of gloomy cold inspire some life-forms to flee the mid-latitude winters, from migrating birds headed south to Arizona's seasonal residents, dubbed "snowbirds" by locals.

Southwestern clear skies look good in winter, but a string of cloudless days in June can bring on an epidemic of eye-rolling among those spending summers in the desert. In August, while northerners are mourning the dog days of summer, southwesterners are celebrating "Monsoon Madness." Festivals, sales, and songs welcome the summer monsoon, with its temporary relief from the high-pressure aridity. Loosely, as the tropical convergence zone moves north with the summer sun, so does the high-pressure descending branch of Hadley circulation. The North American monsoon migrates from Mexico to reach the borders of Arizona and New Mexico within days or weeks after the Southwest hits the longest day of the year. Over the next month, it sidles farther north, system by system, to the Colorado Plateau. As rain clouds cluster in the Four Corners area of the U.S. Southwest, dry skies often expand across the Dakotas, Iowa, and states west and south of them, based on research by Wayne Higgins and others at the National Climate Prediction Center. That's what happens during a strong monsoon, anyway. During a weak monsoon, the pattern doesn't fully shift, leaving the system delivering more rain to the Great Plains and less to the Southwest, according to their analysis of three decades of summer rainfall patterns. Similar seasonal changes in high-pressure systems play out in other regions around the world.

The monsoon illustrates the types of changes in Hadley circulation that global warming may be inspiring to varying degrees throughout the year. Climatologist Andrew Ellis suggests that subtropical air "balloons" up during the monsoon, pushing the high-pressure cell aloft. "It just opens the door for some very light flow and accompanying moisture from the south," he explained during a telephone conversation. When that ceiling lifts, air parcels can float higher in the atmosphere, sometimes reaching heights cold enough to condense vapor into liquid. This invisible ceiling sometimes reveals its presence in the form of anvil clouds. The tops of anvil clouds flatten as they reach the ceiling, more technically known as the tropopause, that divides the lower layer of the atmosphere from the overlying stratosphere. (In reality, the tropopause is more of a transition layer that slows rising clouds than a ceiling they bump against.) Above the tropopause, the warmth of the stratospheric ozone layer—remember, ozone is a potent greenhouse gas—quashes the movement of ascending air, sometimes flattening the tops of clouds. Even where no clouds are forming, researchers can pinpoint the layer's location by the temperature change. Researchers Dian Seidel and William Randel did so using data from weather balloons released around the world. Unlike other regions with relatively stable ceilings, the tropopause in the subtropics shifts from winter lows of about 8 miles to summer highs of about 10 miles, they found. The increase in vertical height basically reflects the horizontal expansion of the tropical belt.

Climate models had predicted an expansion of the tropical belt—although, like glacial melting and hurricane strength, it seems to be manifesting much faster than had been projected. One theory predicting the shift in Hadley circulation goes like this: As tropical seas warm, the water cycle becomes more invigorated. Air rises higher in the atmosphere, carrying its load of former sea spray and future rainfall farther up. As the energized rising branch of the Hadley circulation reaches new heights, its arc swings out farther on the globe. Somewhat as a baseball player flicks extra energy into the ball for a long-distance throw, the extra energy from warming flings the descending branch poleward. Scientists continue to debate theories about the mechanism. In the meantime, global climate models suggest that the descending Hadley branch should shift some 2 degrees poleward by the year 2100, as Seidel noted in a 2008 paper with colleagues. Measurements have confirmed that the descending branch has, in fact, shifted toward the poles. What's more, the shift over the past quarter century apparently has already surpassed that 2-degree projection for the entire century. When Yongyun Hu and Qiang Fu compared five independent research approaches in a 2007 paper, they found that the Northern Hemisphere's descending branch had shifted northward by about 2 to 4.5 degrees, on average, since 1979.

❧

The full implications of this shifting climatic boundary remain to be seen. One of the more ominous predictions involves an expectation for drier conditions in the U.S. Southwest and other subtropical lands, at least outside of the monsoon season. Climate models suggest that the invigorated Hadley circulation increases the risk of extended drought in the subtropics, a region that already features many of the world's deserts. The models generally project that wet regions (such as the humid tropics) will get wetter, whereas dry regions (such as the semi-arid subtropics) will get drier. That's not exactly how things are panning out so far, judging from an analysis of precipitation trends led by Xuebin Zhang and Francis Zwiers, but the warming has only just begun. Zhang and Zwiers separated the globe into horizontal zones stretching across every 10 degrees of latitude and used data from the Global Historical Climatology Network to analyze precipitation from 1925 through 1999. They found an overall decline in rainfall in the areas containing the Northern Hemisphere's humid tropics, from the equator to 20 degrees North. Meanwhile, they found a slight increase in precipitation in the region encompassing the northern subtropics, from about 20 to 40 degrees North. In the other areas of the globe, precipitation generally remained stable or increased, including in the southern subtropics, their analysis found. But precipitation is notoriously difficult to record accurately, much less predict.

Jagged variability turns up in virtually any record of rainfall, with spikes and dips fluctuating across seasons, years, and decades, so it's tough to tell whether precipitation is increasing or decreasing over the years. As a rule, it's much more difficult to measure and model rainfall than temperature. Every spot on the planet has a temperature, day or night, every minute of every hour. It rises and falls in a relatively predictable fashion daily, seasonally, and in response to global warming. Precipitation, on the other hand, is not there—and then, suddenly, it's there. In other words, it's nonlinear. There's a shifting border around every storm, a moving target that can drop rain on one side of the street and not the other. Meanwhile, official weather stations often are cities apart. Even the devices at official stations are subject to the whims of weather. Wind can carry extra rain on a slant into measuring devices or blow some previously entrapped water back out. In severe storms such as hurricanes, the winds can carry off measuring devices altogether. Capturing snow involves equipment only slightly fancier than coffee cans. Then there's the issue of converting snow into water equivalent, with 1 inch of water derived from anywhere between about 4 and 40 inches of snow.

The challenge in measuring precipitation trends to detect whether it's increasing or declining comes through in a chapter by Trenberth and others in the 2007 Intergovernmental Panel on Climate Change (IPCC) report. Data from six different analytical approaches went both ways, with some showing a slight decrease and others showing a slight increase in global rainfall over land between 1979 and 2005. Fewer approaches looked at longer records, but again results came out on both sides of zero. In short, the instrumental record of precipitation varies so much that it's often unclear whether there's been any trend in precipitation at all. The starting point really matters in records of precipitation—the same data can show an upward trend from a mid-century drought or a downward trend from a mid-1970s wet period. In part, this variability results because rainfall patterns tend to shift not only by seasons and years, but also by decades. Even the natural variability that can operate across decades provides reason enough for subtropical regions in particular to prepare for lengthy droughts. Adding higher temperatures—with their higher evaporation rates—increases the potential for severe droughts, whether the drying relates to climate change or natural variability.

Somewhat as Hadley circulation affects where precipitation falls between the equator and the poles, other climate features tend to hold highs and lows in place horizontally across the globe for weeks, months, or longer. For instance, the Bermuda High off Florida can head off hurricanes, thus dictating whether these cyclones enter the Gulf of Mexico or detour north toward the East Coast. As with rainfall in general, longer-term patterns generally relate back to sea-surface temperature. The so-called El Niño climate pattern traces back to sea-surface temperatures in the tropical Pacific. Cutting down the number of Atlantic hurricanes, as mentioned in the first chapter, is only one of its many influences. El Niño and a related longer-term pattern known as the Pacific Decadal Oscillation help determine where rain falls around the globe. Where seas run relatively warm, lows arise that inspire precipitation around and downwind of them. So the balmy Gulf of Mexico helps override the effect of Hadley circulation on nearby states in the U.S. Southeast. Similarly, where relatively cool seas prevail, sinking air and associated high-pressure zones limit cloud activity. The relatively cool seas off the California coast assist Hadley circulation in deflecting rain clouds from the U.S. Southwest. It's the temperature differences, rather than temperature itself, that drive some of the horizontal and meandering air-circulation patterns.

Temperature differences around the globe always exist to some degree, so year-to-year variability will always be a part of the climate mix. When the years add up into tendencies that extend for decades, local people can face problems. Tree-ring records suggest that a decades-long medieval drought helped push the ancestors of the Pueblos (known to some as the Anasazi) out of New Mexico's Chaco Canyon in the thirteenth century. When conditions change at the scale of centuries, though, those changes reverberate across the land. They show up in a variety of records that carry less detail than tree rings, which record annual variations in climate via growth rings. Records reconstructed from ice cores, sediments, and other geological deposits can reveal climate patterns that held at the scale of centuries and millennia. And the piecing together of these records carried some surprises for climatologists considering rainfall rates during the ice ages.

❧

Harsh winds, whipping dust, shifting sands—these terms described conditions on much of the ice-age world. Dust levels around polar plateaus registered about ten times higher during glacial times than during interglacial warm periods, based on ice cores from Greenland and Antarctica. Dust swirled five times more often into the Indian Ocean, judging from sediment cores. Ice cores from Peru's highest mountain suggested that ice-age air held several times more dust during the last ice age than it does now, which indicates that cold, dry winds swept down from glaciers. In the tropical lowlands, blowing dust and sands regularly rearranged sand dunes. Shifting dunes covered vast portions of tropical Africa and Australia during the height of the last ice age. Some of the excess dust could have come from stronger winds. But dryness prevailed over much of the planet, despite the lower evaporation rates that come with cooler climates.

This surprised the climate community. By the twentieth century, everyone knew that glacial periods had been cold, but many scientists had figured that glacial climates would be wetter than interglacial warm periods. "In the sixties, everybody thought if there's ice, that's water, and so it must be wet," recalled Owen Davis, whose youthful appearance and joking manner make it easy to forget that he's been researching artifacts of ancient climates since the 1960s, mainly with the University of Arizona. "It took a while for people to think about ice ages as being drier." After a group of French scientists unearthed ice-age sand dunes under modern rain forests in the Congo in the 1960s, scientists began to consider this puzzling reality. Once James Hays, John Imbrie, and Nicholas Shackleton had pieced together ice-age ups and downs from sea-sediment cores by the mid-1970s, other researchers could hang their own climate findings on their well-dated framework. In time, the evidence grew irrefutable. Ice ages had been drier than interglacial warm periods, with about half the global precipitation of our modern interglacial climate.

Perhaps the misconception arose in part because it was wetter in the U.S. Southwest at the height of the last ice age. Arizona and New Mexico are notorious for serving up a variety of climate indicators, with their barren hillsides that preserve information and expose it for geologists to mine. The evidence made it clear that huge lakes had thrived in the Southwest while glaciers ruled North America. Lake Bonneville engulfed much of western Nevada, while Lake Lahonta covered a huge portion of western Utah. The North American subtropics had become one of the wettest places on Earth.

Precipitation gains in the subtropics were overwhelmed by losses in the temperate, polar, and tropical regions, though. It seemed the subtropics were receiving the precipitation usually reserved for the mid-latitude temperate zone, which then lay underneath tons of water—the frozen kind. In the eastern half of the United States that was not buried by ice, precipitation dropped by roughly a third, researchers led by Colin Prentice inferred from the vegetation changes that occurred. Around the poles, snowfall dropped by about half. In what are now the humid tropics, ice-age rainfall also halved overall, with declines ranging from about a third to two-thirds of modern amounts. That range covers tropical South America, Asia, Australia, and Africa, as summarized in a 1996 review paper by Michael Thomas and Martin Thorp. While Africa's modern sand dunes occupy a relatively narrow region centering on about 25 degrees in latitude, its ice-age dunes covered a somewhat wider area centering on about 15 degrees. Zones of shifting sand dunes extended from the north down into land occupied by Africa's modern-day Mauritania, Niger, and Chad, encompassing Botswana, Zambia, and even parts of the Congo. Clearly, aridity laid a heavier hand on land near the equator during glacial times.

It was as though Hadley circulation was sputtering so mildly that its drought-inducing descent landed on the tropics rather than the subtropics. This would conform to the theory that warming temperatures invigorate Hadley circulation, with a higher lift at its origin, boosting tropical rainfall tallies and propelling the drying descending arc farther poleward with the extra energy. The drier tropics and wetter subtropics suggest a less vigorous Hadley circulation in general during the last ice age.

In contrast, Hadley circulation appears to have operated quite differently during an earlier hothouse, the Eocene. After analyzing the chemistry of fossil wood from trees growing around the poles during this warm period, researcher Hope Jahren and a colleague proposed in a 2002 paper that summer rainstorms were stretching all the way to the Arctic. Other researchers have suggested that the tropical convergence zone, which currently hovers around the equator, migrated up to about 25 degrees North during this hothouse. That might help explain how tropical rainfall could have made its way up to the Arctic during summer.

If these findings hold up to further investigation, they suggest that Hadley circulation shot so high or was somehow buoyed up so that it never fully descended in a drought-inducing way. Researchers so far have found no real evidence of major deserts during the early Eocene. In contrast, they've found abundant evidence for humid climates across much of the globe. "There's a lack of desert data. Does that mean a lack of desert? We simply don't have a lot of data for the tropics, especially on land," said Lisa Sloan, a paleoclimatologist who has focused on the Eocene and other hothouses. In other words, absence of evidence doesn't necessarily translate to evidence of absence. Still, there's hard evidence of salt flats and other signs of aridity from even earlier times, so it's likely that something would have turned up if deserts were widespread during the Eocene. At any rate, it's clear that the poles—currently arid regions—featured higher precipitation rates in both the Eocene and another hothouse period, the Cretaceous. Peter Skelton and the other authors of an informative and beautifully illustrated 2003 book, *The Cretaceous World*, note that the plentiful polar precipitation during this hothouse period suggests the absence of the modern high-pressure cell over the poles.

In retrospect, it makes sense that cold periods run dry and warm periods deliver more moisture, given atmospheric physics. On a warm world that is 70 percent water—and even more during hothouse periods—the physical law that higher air temperatures boost air moisture is bound to wield an important influence. And evidence indicates it did during at least the hothouse time frames considered here, the early Eocene and mid-Cretaceous. Measuring precipitation poses enough challenges in modern times, and there's no way to assess directly rainfall levels of the past. But several lines of indirect evidence indicate the temperature rise and associated changes that came with these hothouses spurred on higher precipitation rates. The rate at which solid rock weathered into clay was much higher during the early Eocene, some 55 million years ago. A similar situation holds for the mid-Cretaceous hothouse of 100 million years ago. This boost in the weathering rate suggests higher precipitation rates during these hothouse periods (as chapter 7 describes in more detail). The type of vegetation thriving in both time frames similarly implies higher rainfall rates, as the next chapter covers.

At its most basic, the higher temperatures of these hothouse climates alone suggest a vastly invigorated rainfall cycle. The amount of water vapor in the air would roughly double for an 18-degree-Fahrenheit rise in global temperature beyond today's norm, as the authors of *The Cretaceous World* note. They report that global temperature averaged as much as 27 degrees Fahrenheit above today's levels during expanses of this period, putting it at a balmy 85 degrees Fahrenheit. That suggests rainfall would have more than doubled during the mid-Cretaceous, when temperatures peaked. Rain also would have fallen in more intense events, with random pockets of condensing water vapor pooling into huge raindrops as they PacManned through a saturated atmosphere. Higher air temperatures mean greater air moisture and therefore more—and more intense—storms. So the available, and even missing, evidence supports our understanding that warmer air delivers more precipitation. To consider another aspect of how ancient precipitation fared, we need to look to the sea.

❧

Sea levels rose and fell over the eons, with myriad effects on climate. During ice ages, a greater share of the global water supply hardens into frozen solids, which makes it unavailable to the atmosphere, to plants, and to animals lacking Bunsen burners. As ice ensnares water, shorelines retreat, leaving less surface liquid to evaporate and feed rainfall. Continental shelves shed their layer of water, thus converting seafloor into landscape. Meanwhile, polar and even mid-latitude regions keep much or all of their water frozen year-round.

During hothouse periods, in contrast, ice melts and water covers more land. More expansive seas provide fodder for evaporation, and thus precipitation. During the late Cretaceous, water covered roughly 76 percent of the globe's area rather than the 70 percent it claims today. A similar ratio would have held in the mid-Cretaceous and early Eocene, given the general lack of sea-level ice throughout both hothouse periods. Few deny that a sea-level rise would prove disastrous for modern coastal lands and cities, with the extent of the damage directly related to the extent of the rise. What is mentioned less often is the impact that rising or falling seas have on precipitation and temperature. Understanding some of the basic climate parameters associated with these ups and downs will enlighten attempts to understand how Gaia handles the heat.

Rising seas promote rainfall and launch more floods. Storm surges can reach further inland. This surely affected coastal areas during hothouse climates. Shallow seas invaded the interiors of continents, too. During the early Eocene, the large lakes stretching across parts of Colorado, Wyoming, and Utah covered what we now call the Green River Formation, a source of fossils and oil shales today. During part of the Cretaceous, an inland sea stretched from the Gulf of Mexico to the Arctic Sea, inundating much of modern-day Wyoming and lands north and south. Where seas invaded continents and lakes spread over the land during the Cretaceous and Eocene, these water bodies often kept climate more humid than their landlocked counterparts. This "lake effect" promotes local precipitation even while moderating wide swings in temperature. (This doesn't hold for water bodies that freeze during winter, though.) With the "continental effect," the farther away a continent's interior gets from relevant water bodies, the more likely it is to be dry, and the bigger the difference between summer and winter temperatures. During warm times of high seas, the lake effect often overpowers the continental effect.

Thanks in part to higher seas, the mid-Cretaceous also featured vast river systems and abundant wetlands. Sea level influences the height of the water table, which in turn determines where and when rivers flow. Across the land, rivers show up where the water table breaks through to the surface. If the water table drops, rivers can move underground. They join the subterranean watercourses. At the surface, it may seem as though a formerly lush riparian area has gone to hell. Tucson has many modern examples of how rivers can dry up when the water table sinks—in this case, from a history of agriculture and population growth that involved extensive pumping of groundwater. Tucson's once-perennial Santa Cruz River now flows only sporadically, during monsoons or long winter rains. The Rillito River acts much the same. Even the San Pedro River, one of the nation's most biologically diverse areas because it attracts migrating birds from throughout North, Central, and South America, has been fading fast in recent years. In 2005, the San Pedro registered zero flow at its Charleston gauge near Sierra Vista for the first time since recordkeeping began in 1930. Environmentalists point to the water demands of the burgeoning local population, which in the 2000s was rising at a rate that would double it in less than twelve years. Developers argue the years of drought that preceded the measurement caused the river's decline. Whatever the cause, and it's likely a mixture of both, the effect is the same: A water table that drops below the surface takes the river down with it. There's no controversy among scientists that the height of a local water table defines where rivers flow. Scientists also readily acknowledge that water tables tend to rise and fall with global sea level, along with local factors. Thus, rising seas bring more water to the surface even on land as water tables follow their lead.

Along with higher seas, the higher sea-surface temperatures of a warmer world would conspire to make it wetter. Where the ocean advanced along coasts and formed inland lakes and seas, underlying land would rest relatively near the surface. These shallow seas, including those on the continental shelves, would warm more readily than deeper seas. In contrast, glacial seas would contract into deeper parts of the ocean basins, largely abandoning continental shelves. The seasonally melting ice and greater proportion of deep to shallow water during ice ages would limit the seas' ability to warm.

Even tropical seas cooled during the ice age, with surface temperatures dropping by about 3 to 5 degrees Fahrenheit. That's a reasonable average considering changes in oxygen isotopes and in the range of warm-water plankton species, suggested Wallace Broecker, an esteemed Columbia University paleoclimatologist, in his self-published 2002 book, *The Glacial World according to Wally*. Local cooling of about 9 degrees Fahrenheit occurred in parts of the equatorial oceans, such as near Barbados and Vanuatu, he noted. These cooler seas would have contributed to the decline in precipitation that registered during glacial climes. Cooler tropical seas and icy northern seas brought down land temperatures, too. During the last ice age, Ireland's temperature ran more like modern-day Alaska's. In modern times, the Gulf Stream carries warm waters from the tropics up to Ireland, England, and nearby lands, keeping them warmer during winter than areas across the Atlantic at similar latitudes, such as Maine. This circulation pattern slowed down during the last ice age.

In the past, Broecker raised concerns that a similar switch could occur in response to warming, but he has since backed away from this prognosis. In the 1990s, he suggested that the melting Arctic ice could shut down the Gulf current, thus potentially flipping Europe into an ice age, with repercussions elsewhere as well. Upon reviewing more evidence, Broecker later concluded that a flip into an ice age seemed unlikely given that the climate first would have to warm by several degrees Fahrenheit or more; at that point, global warming probably would be too entrenched for the Atlantic to shift the climate into glacial conditions. "As I am the one who first raised the warning signal, it is a relief to be able to declare that it may have been a false alarm," he wrote in 2000, in a manuscript he shared with Mark Bowen, author of the 2005 book *Thin Ice*. Nonetheless, the concept continues to thread through the popular culture, with adaptations to suit Hollywood. For instance, the 2004 movie *The Day after Tomorrow* featured a freezing-over of New York in one fictional storm. Although unexpected consequences could stem from global warming, most scientists, including Broecker, generally agree that a global cooling is unlikely to be among the climate surprises that result from the buildup of greenhouse gases in the atmosphere.

❦

While the ice age seasonally turned the Green Isle white, the hothouse Eocene greened lands at the latitude of modern-day Iceland. Even lands at the South Pole favored verdant hues in hothouse climes. Antarctica remained generally free of permanent ice during the Eocene and Cretaceous hothouse periods, as well as several earlier ones. During the early Eocene, arguably the hottest peak of the past 65 million years, polar deep-sea temperatures registered about 18 degrees Fahrenheit higher than modern bottom waters, according to a compilation of data analyzed by James Zachos and colleagues, including Lisa Sloan. Today they hover just a few degrees above freezing. At the surface, ocean temperatures around the North Pole rose by about 14 degrees Fahrenheit, mimicking subtropical climes. Crocodiles followed the heat, paddling up to the latitude of modern-day Oregon, with a few even reaching as far as Canada's Ellesmere Island, near Greenland. Today, frigid meltwater and floating ice keep Arctic seas cool enough (at the moment, anyway) that cold-loving polar bears and penguins swim where crocodiles once ranged. Modern-day crocodiles and alligators draw the line much farther south, about a third of a world away from their Eocene range.

The hothouse seas likely stimulated tropical convection in a big way. As mentioned earlier, there's some evidence suggesting that hothouse climates exported tropical rainfall all the way from the equator to the polar regions. Hurricanes likely strengthened during hothouses and weakened during ice ages, as chapter 1 described. Could monsoons and tropical thunderstorms in general also have worked overtime to transport heat out of the tropics?

Warm temperatures activate summer monsoons in the present, pushing the tropical envelope poleward. Warmer summers in the past generally spurred on monsoon activity, too. It's not clear whether the temperature rise from greenhouse gases will have the same effect on monsoon activity as the earlier warmings from Milankovitch solar cycles that increased the amount of sunlight striking the edge of the Arctic Circle. Still, the warming from the latter did trigger stronger monsoons, judging from the best-studied system—the one that periodically floods parts of India and other Asian nations. The strength of the Asian monsoon waxed and waned with solar power and temperature over the past few million years, based on fragments of information extracted from seafloor muds from the last couple of million years.

Like hurricanes, monsoons often bring bursts of local productivity in affected oceans. These changes leave their mark in ocean sediments. Research linking productivity to monsoon strength spans hundreds or even millions of years, depending on the sediment core examined. A group led by Anil Gupta looked at a well-preserved core from the Arabian Sea near Oman to reconstruct the waxing and waning of the Asian monsoon since the last ice age. Like others before them, they found that the proportion of a key plankton species—*Globigerina bulloides*, which only shows up in the tropics when strong winds create an upwelling of nutrients from deep waters—strongly reflected the peaks in solar warming at high latitudes. In other words, the monsoon strengthened when the Arctic Circle warmed, at least when years are lumped together into average conditions.

Similarly, a research team led by Hartmut Schultz examined the variations in the amount of carbon reaching Arabian Sea sediments over the past 110,000 years. They, too, found that the intensity of the Asian monsoon's productivity, represented by relatively high carbon content in sediments, closely mirrored the ups and downs of Arctic Circle warming and cooling. Particularly weak monsoons often coincided with a northern cooling so severe it rafted chunks of ice out to sea. These episodes of ice-rafting, known as Heinrich events, clearly offer a means for cooling the seas beyond the poles. That ice would melt once it drifted far enough from the poles, cooling local seas like an ice cube in a Pepsi. In fact, scientists recognize these events by the debris they deposit on the seafloor far from the poles upon melting. Cool seas would help dampen the activity of tropical storms—including those arriving with monsoons and hurricanes.

Modern observations indicate that warmer sea-surface temperatures make conditions ripe for a variety of tropical storms, including monsoons and lone thunderstorms as well as hurricanes. Ocean sediments indicate that the Asian monsoon waxed and waned with temperature swings of the geological past. In the past few years, research has pointed to 80 degrees Fahrenheit as a critical sea-surface temperature threshold for a variety of tropical storms, and about 83 degrees for more intense storms. As oceans register a delayed response to rising atmospheric temperatures, their surface layers pass this critical threshold more often.

This is one of many ways a warmer climate can increase precipitation rates. Rainfall rates generally rise as temperature does, as evidenced in modern-day measurements as well as geological history. Evidence spanning ice ages to ancient hothouses makes clear that global precipitation rates increased as climate warmed. High air temperatures and high seas conspire to raise air moisture and thus rainfall rates. Slightly higher temperatures can even boost snowfall rates in some cases, such as around unfrozen lakes and other water bodies. During warm climates, the pouring rains and flooding would help boost another Gaian response to changing climate: plant growth.

5

An Herbal Remedy

Plants Work to Restore Balance

Walking around the fields of winter wheat, I could tell which plants were growing under a futuristic atmosphere. The wheat plants exposed to the higher carbon dioxide levels stood knee high, while their same-age counterparts growing under normal air barely grazed the middle of my shins. None of these plants would end up in Wonder Bread or Ritz crackers. The eight circular fields south of Phoenix were part of a Free-Air Carbon Dioxide Experiment, one of many so-called FACE projects scattered throughout the country. To the casual observer, a FACE plot might look like a field or forest encircled by plastic pipes. But these plain white pipes linked to sophisticated equipment. Monitors would sense the wind direction and strength. Then valves would release just the amount of carbon dioxide needed to keep the air above the test plants carrying about 550 parts per million, the levels projected for mid-century. The collective FACE results confirmed what many indoor experiments had already shown—plants grow faster and bigger under the higher carbon dioxide levels projected for the near future.

Carbon dioxide acts as a greenhouse gas, warming the Earth, as evidence at a variety of time frames indicates. But it plays another important role as well. It provides the basic fodder for plant life. Carbon dioxide launches the chain reaction that provides food for all the world's animals, including humans. Energized by sunlight, plants recombine carbon dioxide with water to create carbohydrates, the starches and sugars we all know and love. Then they use these carbohydrates to add bulk. Basically, plants create mass from energy and two simple ingredients: water and carbon dioxide. The recipe also includes pinches of nitrogen, phosphorus, and other fertilizing nutrients. The equivalent of plant vitamins, these nutrients reach the plant via rainfall or soil water—so water is at the root of even nutrient intake. Thus, the collective mass of plants in a given place and time largely reflects the availability of carbon dioxide, water, and sunlight—along with a factor related to all of them, temperature.

Along with precipitation rates, temperature defines where and when plants grow. In the recipe for plants, it makes a difference whether they bake under the hot desert sun, simmer in the steamy tropics, or barely thaw out from an Arctic winter. Cool climes limit plant growth, partly by affecting whether water exists as a useful liquid or as unavailable ice. The relative lack of sunshine that keeps these climes cool also affects plants' ability to flourish, of course. On the other end, too much sunlight and heat can stress plants to the breaking point, especially when they lack enough water to deal with high evaporation rates. Temperature affects the form precipitation takes—rain, snow, fog—and where and when it falls (as the previous chapter illustrated). All these factors affect how lush plants get and for how much of the year. In turn, the factors that make plants grow—carbon dioxide, water, sunlight, temperature, and nutrients—influence how much carbon dioxide stays in the air to warm the planet.

The doubling of carbon dioxide considered by FACE experiments would pose many threats to the world's people, and to the global economy, because of its projected impact on temperatures. But to consider the atmosphere's composition from a plant's perspective, let's compare apples and oranges. If apples were oxygen and oranges were carbon dioxide, there would be 210,000 apples for every 390 oranges in our existing atmosphere. Pears, representing nitrogen, would comprise most of the other 780,000 parts per million. So plants, restricted to a diet of oranges, can actually struggle to find enough nourishment from modern air. Life is even tougher during ice ages, when carbon dioxide levels dip down below 180 parts per million. As individuals, plants are merely responding to extra resources within their reach when they grow faster under higher carbon dioxide levels. Agricultural scientists taking the plants' point of view have dubbed the effect carbon dioxide fertilization. From the planetary perspective, though, this growth response helps keep levels of this greenhouse gas in check.

While FACE experiments focus on the future, the results also imply that plants grew more vigorously during past warm periods featuring higher carbon dioxide levels. Roughly three-quarters of the last half a billion years of Earth's history involved times when carbon dioxide levels and temperature were higher than modern values. Gaia theory suggests that they would have been even higher without the intervention of plants. Modern measurements confirm that Earth's natural systems have been tempering the current rise in carbon dioxide levels. Could the increase in plant growth that comes with higher carbon dioxide levels and associated changes be one of the ways Gaia regulates its climate?

❧

When wondering whether someone is alive, we first check to see if they're breathing. We can apply a similar concept to the planet. Monthly records of the planet's carbon dioxide levels bring to mind a living, breathing Gaia. In a burst of growth during the northern spring, plants pull carbon dioxide from the atmosphere. They hold it through summer. By mid-fall, as sunlight wanes and growth slows, plants start releasing carbon dioxide. Fallen leaves decay, converting carbohydrates back into carbon dioxide. Growth halts once the ground freezes. The planet's yearlong carbon dioxide cycle reflects plant growth in the Northern Hemisphere, where continents cluster into life-support systems for plants.

The seasonal records of carbon dioxide levels also have helped illustrate something unexpected, though. When the yearlong breath draws to a close in winter, the world's plants are still holding some of the carbon they took in during spring and summer. The living planet is inhaling more carbon dioxide than it is exhaling. More accurately, Gaia's natural systems are storing away some of the greenhouse gases released by human industry, thus slowing the pace of global warming. The missing carbon dioxide inspired an investigation into its fate.

Plants and the oceans retain about half of the carbon dioxide that society releases in a typical year by burning fossil fuels. What's more, plants also capture all of the carbon dioxide released in the burning and destruction of the world's forests—no small number itself. Tropical forest destruction and other land-use changes release about 2 or 3 billion tons of carbon in a typical year, compared to the more than 8 billion tons (and rising) released from fossil fuels. The amount varies from year to year. On average, though, as a 2007 *Science* paper by Britton Stephens and colleagues noted, natural systems are pulling down about 60 percent of the carbon dioxide society places in the atmosphere by burning coal, oil, gas, and forests.

If the world's forests and oceans hadn't been capturing much of the carbon dioxide released since the start of the industrial age, the levels in the air already would have passed 450 parts per million—a goal many scientists view as the place to stabilize if we hope to avoid catastrophic global warming. Stabilizing at 350 would be crucial if we hope to return to the relative climate stability of the recent past, as author and global warming activist Bill McKibben and a growing number of others believe. Carbon dioxide levels were nearing 390 parts per million in early 2010, but given the behind-the-scenes help from the forests and the sea, perhaps there's still hope we can rein in the temperature rise before it becomes a runaway warming. For that, it would help to know the who, what, where, when, and how of the systematic disappearance of carbon dioxide. This chapter and chapter 6 cover the land, while chapter 7 touches on oceanic influences.

❦

A forest has many faces. From a cloud's perspective, it must resemble a series of wicks drawing water from ground to sky. In a bug's world, it must seem a disjoined landscape of surfaces: smooth leaves, rough bark, grainy soil. But from the vantage point of a free-floating molecule, a forest must register as a sea of carbon. Scientists sometimes take that perspective as well, referring to the "carbon pools" or "reservoirs of carbon" in forests. Remove the water sloshing around in the leaves or channeling under the bark and half the remaining mass of trees weighs in as carbon—all of it originally constructed from free-floating carbon dioxide molecules.

Given that carbon dioxide serves as a basic building block of plants, especially trees, some researchers have suggested that boosted growth from carbon dioxide fertilization plays a role in the missing atmospheric carbon dioxide. What's more, the other key factors under investigation in connection with the missing carbon dioxide—nitrogen, water, even higher temperatures—generally respond to rising carbon dioxide levels and associated changes in ways that boost plant growth. The level of synchronicity involved suggests plants have evolved to handle the higher temperatures and evaporation rates that come with higher levels of carbon dioxide.

Countless studies, including hundreds of FACE experiments like the one described above, have pointed to the fact that plants grow faster and often bigger under higher carbon dioxide levels. Trees and other woody plants typically grow 25 to 30 percent bigger in the same time frame under doubled carbon dioxide levels, noted Bruce Kimball, an agricultural scientist who helped design the original FACE projects, including the one in Arizona. Extra carbon dioxide also bumps up the "optimum" temperature for plant growth, as Richard Norby and colleagues documented, meaning plants can grow under warmer conditions. Additional carbon dioxide allows plants to scrimp on certain nutrients, and it reduces the amount of water plants need to create a specific amount of carbohydrates, thus improving their water-use efficiency.

The improved water-use efficiency by plants growing under higher carbon dioxide levels even comes with a clear mechanism. Namely, water escapes a plant from the same place carbon dioxide enters. When these gateways, known as stomata, open to let carbon dioxide in, they inadvertently also give water molecules a chance to slip out. When carbon dioxide abounds, plants can get their fill faster, then quickly shut those stomatal gates to keep their water reserves behind closed doors. Cactuses can thrive in conditions that would turn most plants into shriveled stumps because they know how to keep their stomata shut. They wait until night falls to collect most of the carbon dioxide they will need for their modest efforts at photosynthesis.

As with water, plants use nitrogen more efficiently under elevated carbon dioxide levels. Thus, plants are less hampered by nutrient-poor soils or dry spells when carbon dioxide levels run high. The more efficient approach likely comes with a downside, at least from the perspective of humans and other animals living in an elevated carbon dioxide environment. Nitrogen is a key ingredient in proteins. So less nitrogen use means fewer proteins tucked in amid those starchy carbohydrates—an issue that could affect species up the food chain, including us. In fact, when researchers considered how cereals and bread might fare by actually milling some of the wheat grown in a FACE experiment, they found 14 percent less protein in the flour made from plants grown under elevated carbon dioxide conditions compared to those grown in our current atmosphere. Down the line, that could translate into less protein in bread. Another study found that calcium and zinc levels were 10 to 20 percent lower in Illinois soybeans grown under a high carbon dioxide FACE plot compared to those grown under normal conditions.

Higher temperatures can collaborate with carbon dioxide to boost annual growth by expanding the growing season in cool climates or at high elevations. After examining annual satellite images of Northern Hemisphere plant growth, Ranga Myneni and others estimated that the northern growing season came progressively earlier during the 1980s, with an overall increase of about twelve days throughout the decade. They attributed about two-thirds of the expanded growing season to an earlier spring green-up, and the rest to a later fall. The results mirror on-the-ground research showing that growing seasons have lengthened.

Terry Root and her colleagues at Stanford University analyzed 145 records of blooms, green-up, and other seasonal signs of spring. The records, mainly from Europe but also Asia and North America, showed that trees were launching their spring growth about five days earlier since record-keeping started, about twenty-eight years ago on average.

The growth of trees and other plants is expanding in space as well as time. Seedlings are moving up mountains and closer to the poles, beyond the typical "treeline" cutoff. In alpine forests, seedling survival often requires an unusually warm summer—or a series of them. By growing where no trees have grown before in this modern climate, they are expanding the area where vegetation is bringing carbon down to earth. In addition, mature trees living near the treeline border have expanded their annual growth compared to previous years. Researchers also have documented an ongoing surge in growth patterns of many long-lived trees, such as the bristlecone pines that can cling to a precarious existence on alpine slopes for thousands of years.

With bristlecone pine and many temperate trees, it's fairly straightforward to consider growth patterns going back hundreds, even thousands, of years by measuring annual growth rings. Tree rings delineate every year's growth because winter's approach leaves a trail of cells of dwindling size that look noticeably darker than the large cells that announce the arrival of spring. By taking pencil-sized cores and cross sections from thousands of trees, stumps, and long-dead logs, a variety of researchers have documented increases in annual ring width over the past century. Even more consistently, many trees growing near treeline boundaries have put on more wood annually since about 1850, based on the estimated area of a year of growth around the trunk. A consistent climb in this measure of growth leveled off around the middle of the twentieth century at some sites, but growth remains impressive.

Since at least 1984, when several LTRR researchers led by Valmore LaMarche published a *Science* paper on the topic, tree-ring scientists have been debating how much of the growth gain comes directly from the temperature increase, or whether carbon dioxide fertilization also wields an influence. Bristlecone and other mountainside trees typically deal with short growing seasons, often with only six or eight weeks of good weather between chilling frosts. They also face the thinner atmosphere found on high peaks. Much as mountain climbers struggle to get enough oxygen once they reach certain heights, high-altitude trees can face limits from a rarity of carbon dioxide. A bit extra in the air can help them out. One elegant 1991 study by Lisa Graumlich, now the head of the University of Arizona's School of Natural Resources and the Environment, worked to untangle this problem by examining five tree-ring records from California's Sierra Nevada mountains. She found that growth patterns for three of the records generally matched the variability in climate. The other two, meanwhile, demonstrated recent growth above and beyond what would be expected based on climatic conditions, suggesting that carbon dioxide fertilization might have contributed to the growth trend. To create long-term temperature records, tree-ring experts work hard to avoid situations that reflect conditions other than temperature. But for our purposes here—considering how forests fare with global warming—the distinction between the two forces fades in importance. Increases in both temperature and carbon dioxide encourage many temperate trees to put on more wood, and thus lock up more carbon. What's more, the higher temperatures that come with higher carbon dioxide levels can promote growth in other ways as well.

As soils warm under trees living in cold climates, they can release more essential ingredients of growth—not only water, but also nitrogen. Sune Linder and colleagues including Paul Jarvis heated up near-surface soils in a Sweden spruce forest by 9 degrees Fahrenheit throughout the growing season. After five years of the experiment, they found that trees above the warmer soils had grown more than half as much again as their neighbors on the soils left unheated, as Jarvis and Linder noted in a *Nature* brief. The researchers attributed the extra growth in part to the increased availability of nitrogen in the warmer soils. So there's another reason to expect forests in cool regions to respond to global warming by growing better.

Nitrogen levels have increased for reasons unrelated to temperature, too. Nitrogen intended to fertilize crops may float downstream or downwind, to other parts of the landscape. Some of it becomes airborne, to be brought down to earth later by falling raindrops. A nitrogen derivative, nitrate, is one of the compounds making rainfall more acidic in areas around and downwind of farms and industry. Acid rain can harm lifeforms in systems already on the edge, such as soils and lakes without nearby buffering from limestone and other sources of alkalinity. To systems with alkaline buffering, the nitrates and sulfates that form acid rain often act as fertilizers. In water, an abundance of nitrogen boosts plant growth, as when sewage feeds algal blooms. At the global scale, the extra nitrogen floating around the atmosphere and in the rivers likely is boosting overall plant growth (including undesirable algal blooms in lakes and coastal areas). A 2007 journal article by Federico Magnani and others suggests that the extra nitrogen could help explain why the world's vegetation is taking up more carbon than expected.

A similar case has been made for phosphorus, another fertilizer applied to crops. Tropical forests in particular tend to benefit from phosphorus input. In fact, burning tropical forests release phosphorus and nitrogen. The nutrient release partly explains why indigenous rain-forest farmers developed swidden agriculture in the first place. The practice is more commonly known as "slash-and-burn" in these days of burgeoning populations and industrial-scale logging and ranching. Even so, the nitrogen and phosphorus "fertilizer" lingering in the ashes of former trees might help explain how the remaining tropical forests are managing to take up most of the carbon released by their burning brethren, the stands subject to slash-and-burn.

For that matter, extra rainfall could well be providing a boost to plant growth. Global warming brings an overall increase in precipitation (as the previous chapter described). In our current environment, water ranks as among the most important of the essential ingredients for plant growth. Just add water, and plants will pop up. In the moist tropics of Puerto Rico, 6-foot-tall trees can sprout from the sides of buildings, or on top of rocks. Where water can trickle in, flowers can sprout from cracks in the sidewalks that entomb the native prairies of Illinois. In the Sonoran Desert of Arizona, grasses sprout from seemingly barren soils within days of monsoonal rains. In desert areas, though, the growth spurts tend to be rather ephemeral. Subsequent dry spells can kill or burn off some of these plants. Where rainfall levels increase on a relatively consistent basis, as is occurring in much of the eastern United States, plants may grow more enthusiastically, plucking carbon dioxide out of the air as they do.

Measurements of forests in fifty-five nontropical countries showed that mature northern forests were also taking up carbon—enough to account for about a sixth of the world's annual fossil fuel emissions. Well-measured northern European forests were growing bigger, increasing their biomass. So were forests in Mongolia, tree-ring records revealed. What's more, trees around the world were establishing in places once too cold to support them, as described earlier. They've been creeping up mountains and edging toward the poles. Forests and shrubs around the world also are responding to the changing environment, pulling down carbon dioxide in the process.

Trees are even invading grasslands. A variety of shrubs have been expanding into deserts and grasslands across the country. Creosote is invading grasslands in the New Mexico desert. Shrubs are overtaking Wyoming sagebrush. Mesquite is encroaching upon Texas savanna. Dogwood is overshadowing Kansas tallgrass prairie. Across the country, and especially in the West, shrubs and other woody plants have claimed turf on somewhere between 540 million and 800 million acres. Again, the cause of the change is unclear. Besides climate and carbon dioxide, causes could involve grazing pressure or a lack of fire in grassland systems, among other things. But, whatever the cause, the effect is a greater uptake of carbon dioxide. A study by Richard Houghton and colleagues estimated that this so-called woody invasion could account for up to a third of North America's extra uptake of carbon. Other researchers, such as a team led by Robert Jackson, have challenged the concept that trees and shrubs take up more carbon than grasslands. Continuous grasslands can outpace some woodlands, particularly in the desert. Overall, though, the invasion of trees and shrubs increases carbon-collecting productivity, concluded a team of researchers, led by Alan Knapp, comparing results from eight different U.S. sites. On average, the invading shrubs had quadrupled the sites' productivity compared to the original grasses, these researchers reported in a 2008 paper. Of the ecosystems Knapp and his colleagues examined, only desert grasslands measured a decline in productivity due to a shrub invasion.

Trees thrive where there is adequate energy to power them, with sunlight and water and carbon dioxide available in adequate amounts. The fact that forests generally require more water than grasslands suggests they capture more carbon. What's more, adding extra carbon dioxide to the system tips the scales in favor of forests. Elevated levels of carbon dioxide consistently favor trees and herbs over grasses, as M. Rebecca Shaw and colleagues concluded in a 2005 literature review. Overall, trees are responding to the changing atmosphere and climate more quickly and enthusiastically than grasses. The analysis by Terry Root and colleagues considering how 145 species of plants changed their leafing and blooming habits over the past three decades found that trees shifted their growing seasons ahead by five days on average, whereas grasses moved up in time by only two days. And where forests flourish, carbon dioxide comes down to earth.

Even mature tropical rain forests seem to be growing better. By the mid-1990s, on-the-ground tallies began to show that the Amazon's tropical rain forests that had escaped destruction were growing unexpectedly fast. Measurements of tree growth from more than a hundred plots in the Amazon's lowlands suggested that mature forests in this region alone could be pulling down about 10 percent of the carbon dioxide emitted from fossil fuels in the 1980s and 1990s, a 1998 *Science* article by Oliver Phillips and others suggested. Other on-the-ground studies, such as one led by John Grace, similarly concluded that Amazon forests, even old growth, were taking up carbon. There's not much information for tropical rain forests outside of the Americas, but the little there is doesn't show much evidence for similar vigorous growth in Africa or Asia, as the analysis by the Phillips team indicated. Still, the results for the Amazon carry major weight on the global scale. South America accounted for about 42 percent of the world's tropical forests in a 2003 assessment by the United Nations Food and Agricultural Organization that also included tropical dry forests. Debate continues, with scientists including longtime tropical forest researcher Deborah Clark challenging the extent of the uptake and questioning the measuring techniques. "It's like hen's teeth, finding the same plots" and measuring them again to check for weight gains, Clark said during a 2009 talk in Tucson, indicating she was referring only to plots that had not been set up specifically for long-term research. In a 2007 review paper, she outlined other issues, including possible differences in methods for measuring biomass and complexities that can include a history of harvesting or disturbance unbeknownst to the researcher. She and some others remain unconvinced that the extra carbon dioxide is boosting growth in tropical forests. Scientists using isotopic analysis have concluded, however, that tropical forests are currently playing a major role in clearing the air of carbon dioxide.

The ground-based studies showing that the world's tropical forests are taking up significant amounts of carbon, for whatever reason, received support in 2007 from a different perspective—the air. After comparing carbon dioxide levels and chemistry from multiple heights in the atmosphere from a dozen sites, Britton Stephens of the National Center for Atmospheric Research and others concluded that tropical forests around the world were capturing virtually all of the carbon emitted via the razing of regional forests. Depending on the year, the carbon released in the burning and destruction of tropical forests amounts to a quarter or more of the carbon released in the burning of fossil fuels. Stephens's co-authors read like a who's who list of researchers involved in tracking down the missing carbon sink using atmospheric chemistry—including several who had earlier theorized that northern forests were the main heroes. So it's convincing that they concluded tropical forests deserved almost an equal share of the credit for uptake of the missing carbon. Their findings are especially interesting given that there's no reason to expect the ongoing temperature rise to benefit tropical forests—they already exist in frost-free conditions. The findings seem less surprising, though, in the context of where carbon dioxide comes to rest when it materializes into carbon on land. As the next section explains, tropical rain forests hold more than their fair share of the carbon locked up in the world's vegetation.

❦

The world's forests contain roughly 80 percent of the biomass, or dry weight, found in global vegetation. The world's trees hold nearly as much carbon as the miles of air drifting above them. That's about 650 billion tons of carbon, estimated a 2004 publication by Nicolas Gruber and colleagues. (Forest soils hold roughly triple that amount of carbon, but that's a topic for the next chapter.) The biomass of living forests comes almost entirely from their plant life. Typically, animals comprise less than 1 percent of a forest's biomass, whereas wood in the trunk, branches, and roots accounts for more than 95 percent.

Tropical forests accounted for roughly 60 percent of the carbon stored in forest vegetation in the mid-1990s, a detailed analysis by Robert Dixon and others found. That's despite the fact that they take up little more than 40 percent of the global area of forests. Meanwhile, the boreal forests of Alaska, Canada, Siberia, and other near-polar regions accounted for about a third of the world's forest area but only about a quarter of the carbon stored in global forests, this analysis found. The forests in the middle—those of the United States, Europe, China, and other countries in the temperate zone—accounted for the remainder, a paltry 16 percent of the world's carbon in forest biomass. This low value partly reflects the large-scale forest clearing that has occurred over centuries and millennia in developed countries, but climate also plays a role.

At some level, we all recognize that local plants reflect the climate in which they grow. If we were watching a movie supposedly set in the tropical rain forest, the presence of bent-arm saguaro cactuses would pull us out of the scene and back into our chairs, onlookers to an illusion. We know the jungle should look lush and verdant, while sparse and spiky belongs in the desert. Measurements from a variety of ecosystems bear out our intuitive understanding that plant growth reflects climate. Leslie Holdridge developed a system in the 1940s using climate to predict the types of plants that would colonize a site. The Holdridge system and others like it focused on average temperature, precipitation, and evaporation rates to predict what should grow in a specific location. These models work because temperature—and the water leftover after evaporation has taken its toll—dictates how many layers of greenery can form, if any. From there, it's reasonable to estimate a vegetation type's dry weight, or biomass.

In general, plant growth reaches an apex of enthusiasm in humid forests. Water is relatively plentiful in the rain forest—hence the name. Add to that sufficient sunlight and warm temperatures that stay above freezing, and you've got a recipe for black forest, trees sporting a canopy so thick it intercepts 98 percent of sunlight before it even hits the ground. The Holdridge system and others like it place tropical rain forests at the top of the heap, ranking them as the most productive of the land-based ecosystems. Clearly, a hot, humid climate can nurture plant growth.

❧

During a summer visit to Puerto Rico, my family wandered near one of Old San Juan's public gardens south of El Morro Fort. "Is that a philodendron?" asked my nephew Aidan, walking over to a vine dangling with the familiar heart-shaped leaves of this popular houseplant. Individual leaves, green laced with pale yellow, stretched from his elbow to his fingertips. "Twelve-foot-tall houseplants!" he marveled. Many Puerto Ricans hold a special place in their hearts for another huge-leaved tropical tree: yagrumo, known to scientists as *Cecropia*. During Navidad, the Christmas holidays celebrated by the Caribbean island's many Catholics, islanders tie red ribbons around the stems of yagrumo leaves and hang them up for decoration. From tip to tip, one leaf can cover a turkey platter. The fast-growing yagrumo trees thrive in open spaces throughout the tropical Americas.

Yagrumo aren't the only tropical plants growing as if they're on steroids. Look at banana-plant leaves. Shaped like the blades of a ceiling fan, the biggest of them could cool down an opera house with their 6-foot spread. Even grasses in the moist tropics grow to heights unimaginable for climates that face annual frosts. Bamboo shoots on this tropical island reach 15 feet tall or more, propped up by a woodlike stem that belies its membership in the grass family.

The sheer mass of leaves sustaining themselves at a site gives a strong indication of its overall productivity. So does the mass of wood. Together, the biomass of leaves and wood tends to reflect productivity. And while tropical rain forests rank among the most productive systems in the world, nothing beats the temperate rain forests of the Pacific Northwest for sheer size. In a 2007 study comparing a variety of forest stands around the world, researchers Helen Keeling and Oliver Phillips calculated that the visible biomass in tropical lowland forests averaged about 700 pounds per square foot, while several stands of coastal redwoods (*Sequoia sempervirens*) assessed weighed five to ten times more—greater than any of the tropical stands measured by scientists. And these averages count the space between trees. The numbers get so high because so do redwoods. Coastal redwoods and giant sequoias both can approach 400 feet tall (although most of the tallest ones have been cut down, as Elliott Norse notes in his book, *Ancient Forests of the Pacific Northwest)*. What's more, a dozen other massive conifer species in the Pacific Northwest can tower above 200 feet.

Why do plants grow so vigorously in the rain forests of the tropics and Pacific Northwest? The answer goes back to the basic ingredients of plant life. Based on modern observations and experiments, we know that plants grow fastest in frost-free climes, where water abounds but doesn't drown plant roots. If carbon dioxide levels are running a bit high, all the better from a plant's perspective. The moist tropics create conditions that support hardwoods, which are angiosperms. They generally outcompete gymnosperms—pines and other softwoods—wherever conditions are conducive to plant growth. The coastal regions of the Pacific Northwest, meanwhile, have climates that can prove more challenging to hardwoods. They have relatively dry summers—especially if you count only rainfall and not fog. A difficult factor to measure, fog nevertheless appears to be crucial in sustaining California's coastal redwoods, which are gymnosperms, especially during otherwise dry summers. The greater surface area per volume of gymnosperm needles compared to angiosperm leaves might aid in collecting drops of fog into nourishing precipitation. Researchers have found that these conifers are, in fact, able to use the fog that comes their way, which can amount to a tenth to half of total water collected by a tree annually, a three-year study by Todd Dawson indicated. He also found that without the trees, the amount of fog accumulating at a site dropped in half.

Coastal redwoods generally thrive in frost-free climates, but some of the other big conifers do face freezing conditions from time to time. Even as far north as Washington State, though, the Pacific Northwest coastal area has fewer freezing nights in January than does northern Louisiana, as Norse reports. A hard frost turns water to ice in the veins of many broadleaf plants, shriveling once-healthy leaves into brown remnants within hours. That's partly why many northern angiosperms drop their leaves in autumn. Conifers, on the other hand, can weather the occasional frost and be ready to photosynthesize as soon as water turns liquid again. Cold-climate evergreens don't reach the heights of the towering giants of the Pacific Northwest, though. In the frigid climes found on temperate mountaintops and in near-polar regions, spruce can grow on shallow soils above permafrost—but they don't grow very big. In the Pacific Northwest, in contrast, abundant winter precipitation and summer fog come with mild temperatures. So the old-growth rain forests of Oregon, Washington, and California amount to the northern equivalent of mature tropical rain forest. Both systems hold a disproportionate amount of the world's biomass considering the area they cover.

The links between plant size, productivity, and biomass help explain why the world's rain forests have come under a magnifying glass of international attention. Forests are key to the carbon dioxide drawdown, with rain forests in the Amazon playing the biggest role. The Amazon is one of the few regions of the world where widespread, highly productive forests still dominate the landscape. In the United States, settlers decimated the temperate rain forests decades ago. A mere 5 percent of the area covered by looming coastal redwoods in northwestern California and giant sequoias in California's Sierra Nevada in the middle nineteenth century remains as primeval forest range. European forests have faced even more development pressure, although an old-growth stand that has survived intact for more than seven centuries suggests that Europe's mid-latitude trees also could grow large, given enough time. The protected old-growth forest straddling Poland and Belarus features 7-foot-wide ash and linden, recalled Alan Weisman, author of the best-selling 2007 book *The World without Us*. "You're looking at a fragment that used to go from Siberia to Ireland," he said at a book-signing event that year. Such fragments have otherwise gone the way of the coastal redwood forest. These days, European companies have to get much of their timber from trees around a foot in diameter, a size most U.S. companies consider too small to harvest. Meanwhile, ample evidence suggests that South America's indigenous people, such as the Mayas and Incas, commonly cleared rain forests for their pre-Columbian civilizations. After disease and conquest wiped out many of these civilizations, industrial-scale deforestation didn't resume until the 1970s in the Amazon. At the start of the new millennium, tropical rain forests still covered close to two-thirds of Brazil's land, although they continue to lose ground every day despite the Brazilian government's protection efforts.

Computer models suggesting that Amazonian forests will lose more ground to climate change regularly make headlines. In fact, modeling studies led by Richard Betts and Peter Cox of the United Kingdom's Hadley Centre apparently served as the main reason for James Lovelock's concern that global warming could deliver a fatal blow to the world's forests and thus to Gaia, judging from his mention of their work in his 2006 book *The Revenge of Gaia*. In a 2000 paper led by Cox and a 2004 paper led by Betts (working with others in both cases), they modeled widespread forest destruction, proposing that regional temperatures would increase by about 18 degrees Fahrenheit while precipitation would drop by 60 percent by century's end because of the conditions they included in their model. One of the conditions harkens back to the documented ability of plants to use less water as carbon dioxide levels rise. With those stomata-shutting plants clinging to their water supplies, fewer water molecules could free themselves to gather into clouds, the researchers speculated. So instead of having the extra carbon dioxide boost plant growth, as hundreds of field experiments have documented, their model projected that the Amazon forest would dry out because plants would circulate less water. In other words, the model created a situation where the trees' ability to conserve water backfired and led to widespread forest-killing drought.

The Hadley Centre's dire prediction garnered the most media attention, but modeling results predicting how plant life will respond to changing conditions splay across a wide spectrum. For a 2006 paper, Pierre Friedlingstein worked with Cox, Betts, and a couple dozen other researchers to review eleven different model projections on how climate change and other factors might combine to influence the amount of carbon dioxide taken up by the world's vegetation and oceans in the future. At one extreme was the Hadley Centre's model, described above, which projected that forests would die back and release about 6 gigatons of carbon a year by century's end. That's approaching the amount currently released by fossil-fuel emissions. At the other end of the spectrum, researchers using a U.S. Lawrence Livermore National Laboratory model concluded that the power of vegetation to pluck carbon from the atmosphere would actually increase, to total about 11 gigatons of carbon a year by century's end. The results from the other nine models ranged in the middle ground, although only one other besides the Hadley model projected that land plants would reverse their long-standing tradition of taking up carbon dioxide. Clearly there's a long way to go when it comes to modeling future forest and carbon dioxide exchanges. As Friedlingstein and his colleagues concluded in their 2006 paper, it would help if the models were constrained by actual data from observational studies. "To do this," they acknowledged, "the models need to be more complete so that they are more obviously comparable to the real world." At the moment, there's no requirement that they be rooted in reality.

Computer models do not offer a crystal ball to gaze into when considering the long-term fate of forests in a warmer climate. The fates of wetlands, grasslands, and deserts also remain difficult to project given our current state of understanding. Gazing into a microscope at pollen collected in ancient sediments, however, just might help clarify how these ecosystems can fare when the going gets hot. Pollen grains in sediments of known dates provide evidence for the victors in the age-old species competition. Among other things, they highlight the fluctuations between grasslands and forests at many sites around the world, including the tropics. The location of the fossil evidence of certain animals, particularly the ectothermic ones we call "cold-blooded," also can help identify a temperature range during past ages. When it comes to sharing information about past climates, though, the plant kingdom reigns.

The huge leaves sported by tropical plants like yagrumo, banana, and bamboo indicate a warm, moist climate. And that's about what would have been predicted in a technique developed by Jack Wolfe, a longtime University of Arizona researcher who spent many years puzzling out how leaf traits reflect climate. Although banana leaves would be off the charts for Wolfe's technique, they fit the general pattern he recognized: Large leaves with smooth margins grow in hot, humid environments. Thanks to all those lush leaves, tropical rain forests win the competition for most carbohydrates produced in a year per square foot of ground. Small leaves, meanwhile, abound in either cold or dry climates—places where plants struggle to find usable water. In semi-arid

Tucson, this is illustrated well by our typical native trees. Mesquite and paloverde trees seem to have a talent for making miniatures. Two or three individual leaves from either species often can fit onto the trimmed fingernail of my pinkie.

After refining his technique by comparing hundreds of leaves' sizes and shapes to the climate that supported them, Wolfe applied this understanding to fossil leaves to interpret past climates. Wolfe's system to gauge local temperature and rainfall by measuring the sizes and shapes of leaves generally works well for the past 100 million years, since the time the flowering plants known as angiosperms began to dominate the plant world. The needle-like leaves of pines and other conifers carry less information, as this effective arrangement can work from the tops of snowbound mountains to the balmy coasts of tropical seas. Beyond size, the raggedness of angiosperm leaves—that is, the proportion of leaf margin to leaf area—says a lot about an area's average annual temperature, now or in the past. Smooth leaves more often occur in warmer climates, whereas ragged margins seem to promote the movement of water through leaves and so turn up in cooler climates that feature lower evaporation rates. For Wolfe's method to work, detailed information on these and other leaf characteristics from a variety of local species is required. Taken together, though, these characteristics can enlighten our understanding of past climates.

The connections between leaf size, climate, and plant productivity are among the techniques making it possible to piece together snapshots of productivity from forests of the past 100 million years. Peat and coal can preserve plant material so well that even microscopic features remain intact for hundreds of millions of years. Leaf structure sheds further clues about ancient climates. Even the timing of the emergence of evolutionary traits can yield information. All of these techniques help when considering how the world's plants responded to climate changes before humans were around to tip the balance.

Today, Alaska's Colville River winds through barren tundra, with brown peaty soils giving way to gray shaley hills. The sparse plant life does little to break the view of the horizon, observers indicate. The view from this North Slope area looked quite different during Cretaceous times, though, as researchers Judith Totman Parrish and Robert Spicer documented. Geologists who focus on ancient climates that existed long before the ice-age/interglacial cycles of the Quaternary began, Parrish and Spicer mined the ancient sediment deposits on the North Slope for clues about climate and life some 100 million years ago during the mid-Cretaceous, reporting their findings in a 1988 paper and subsequent articles and books. Embedded in these deposits of ancient mud known as the Nanushuk Formation, they found logs and branches—even upright tree trunks. Lush, polar forests once thrived in this now-barren locale.

Their finding and similar excursions by others helped establish the undeniable presence of polar forests during the hothouse Cretaceous. Some of the trees reached massive stature. Fossilized trunks from the mid-Cretaceous showed that trees grew as wide as a lion's girth and taller than a giraffe, even at 85 degrees in latitude, mere degrees away from the North Pole. Applying the leaf-analysis technique developed by Jack Wolfe to these near-polar sites, Parrish and Spicer estimated that the area's average annual temperature registered roughly 50 degrees Fahrenheit—about the same as modern-day London. The many tree species that thrived in this thawed-out version of the Arctic included the mutual ancestor to bald cypress and *Metasequoia* (dawn redwood), both of which favor humid climates that rarely face hard freezes. What's more, there's no evidence of frost found in Alaska's well-preserved Nanushuk wood fossils. In a 1998 paper, Parrish, Spicer, and colleagues documented that forests also grew in New Zealand, then located within the Antarctic Circle.

By the mid-Cretaceous, hardwood trees had expanded their territory to include near-polar regions. Hardwoods belong to the colorful clan known as angiosperms, which span everything from sweet-smelling roses to the mighty oak, and every other flowering plant and hardwood tree in existence. While angiosperm diversity exploded around the planet during the Cretaceous, conifers and other gymnosperms dominated cooler regions such as the poles. Cedar, redwood, and dawn redwood (*Metasequoia*) stood tall among these gymnosperms, explained Scott Wing, a researcher and curator with the Smithsonian's National Museum of Natural History. "*Metasequoia* was a very important component of wetland vegetation," Wing said in a phone conversation, noting that these dawn redwoods were ubiquitous in Europe, Greenland, eastern Asia, and other latitudes above 40 degrees.

These looming Cretaceous polar forests pose quite a contrast to today's Arctic environment. These days, the ice, permafrost, and freezing temperatures in the Arctic Circle stunt modern trees. Knee-high willows dot the landscape until trees dwindle and then vanish. "It's generous to call them trees," remarked Parrish. At the extreme, the Arctic alpine snow willow can have leaves just a fifth of an inch long on a woody "trunk" that stands less than half an inch high. Only shallowly rooted plants like these can grow on tundra permafrost, which replaces forests where July temperatures drop below the 50 degrees Fahrenheit needed to melt permafrost—typically at about 70 degrees North these days.

During the Cretaceous, as now, the long-lasting days of June in the Arctic Circle provide about 90 percent of the amount of solar energy reaching Washington, D.C., in the same month. As any visitor to our nation's capital will recall, that's enough sunlight for growing trees—as long as the soil isn't buried under miles of ice. Still, Arctic winters also featured months of twilight and outright darkness, as they do now. How did Cretaceous trees cope? Work by Parrish and Spicer suggests that mid-Cretaceous gymnosperm forests responded to the long, dark winters by shedding their leaves in unison, much as deciduous angiosperm forests do today. Not all scientists agree with the interpretation that polar trees survived dark winters by dropping their leaves. But researchers concur that these polar forests prove that an unusually warm global climate existed when they did, including during the Cretaceous.

With all this heat centered around the poles, you would think the tropics would be almost unbearably hot. This may have been the case on the continents when the Cretaceous began some 144 million years ago, but things shifted in time. In the early Cretaceous, the continents we now recognize as South America and Africa were still fitted together into one piece. Having such a wide expanse of land creates a "continental effect," a phrase that conjures up interior drying and severe temperature swings. Evidence indicates that this megacontinent featured extensive dryness in the tropics, even around the equator, for millions of years on end. Salt deposits on the landscape indicate an arid environment while those two continents clung together. Some dinosaurs did manage to survive in this region, so apparently springs and pools surfaced in crucial places. But as of 2003, *The Cretaceous World* authors, which include Spicer along with lead author Peter Skelton, had unearthed no signs of tropical rain forests during this early warm period except in a few near-coast areas in what is now Colombia and Southeast Asia.

By the mid-Cretaceous, though, after an exceptionally active time volcanically, South America and Africa had started going their separate ways. Their parting changed air and water circulation patterns, in this case for the better. The equatorial region took on the hot but humid conditions common in much of today's tropics. A belt of humidity circled the globe, encompassing what is now the Amazon rain forest. Across the narrow but expanding Atlantic Ocean, this humid belt also took in what is now Africa's Congo region with its looming rain forests. It even covered about half of North Africa's modern-day desert. As the Cretaceous continued, the belt carrying sustaining rains began to thin again. The decreasing size of the tropical humid region coincided with an ongoing drop in temperature from the mid-Cretaceous peak. But it's difficult to say whether there's a connection, given the rough scale of resolution delineated. When describing Cretaceous conditions, researchers often lump together tens of millions of years into one general picture.

It's challenging for researchers to estimate the productivity and biomass of even modern-day systems. Attempts to estimate the productivity and biomass of plants in the past face similar challenges, magnified by the need to project ecosystems based on scattered evidence. With that caveat, it's still worth noting the results of a model simulating mid-Cretaceous conditions and their effects on plant growth. By combining and tweaking existing models, David Beerling of the University of Sheffield and his colleagues developed a way to estimate net primary productivity based on climate parameters, carbon dioxide levels, and other factors such as decay rates and related nutrient release. With this approach, they estimated that land-based plants during the mid-Cretaceous were roughly twice as productive globally during this warm, wet period of Earth's history than they are today. The extra productivity would have boosted plants' ability to draw down airborne carbon dioxide into biomass and other carbon storage systems. As the researchers note, their modeling results do reflect patterns of productivity where evidence exists, such as in fossil tree rings and preserved plants. And their soil carbon results overlap well with known deposits of mid-Cretaceous coals. (Coals form when plants are pulling carbon beneath the surface, a topic of the next chapter.) Their tropical sea-surface temperatures were higher than evidence suggested when they described the results from their model in 1999, but later research by Karen Bice and colleagues bumped up those temperatures.

Elevated carbon dioxide levels may have accounted for about a third of the increase in global productivity, the analysis by Beerling and his colleagues suggested. Their model indicated that the elevated carbon dioxide level had the biggest impact near the equator. They attribute this to the plants' ability to use less water. As mentioned earlier, high carbon dioxide levels allow plants to keep their stomata closed more often, thus giving their internal water fewer opportunities to escape. This provides the biggest advantage where temperatures run high (as explained later in this chapter). For their analysis, they estimated carbon dioxide levels at 1,000 parts per million, about three times higher than early 1970s levels. Their estimates of carbon dioxide levels fall closer to the low end of Cretaceous estimates—and within the range of projections for the reasonably near future unless society manages to rein in emissions. Although Beerling and his colleagues suggest that their productivity results might rest on the high end, they and others who study this era generally agree that mid-Cretaceous plant life thrived at the global scale.

❧

As in the Cretaceous, massive forests dominated polar regions during the early Eocene, which started about 55 million years ago. The lush Eocene plant life around the poles never faced hard frosts, judging from the "nearest living relative" method of assessing climate using fossil plants—which involves considering climatic conditions where their closest kin now thrive. As in the Cretaceous, lands near the Arctic Circle featured *Metasequoia* and other plants that today refuse to grow in climates with deep winter freezes.

"Climatically these are fairly similar times," said Smithsonian curator Scott Wing, referring to the Eocene and mid-Cretaceous hothouses. "It's pretty clear that the poles are relatively quite warm. And there seems to have been relatively low temperature seasonality" even in landlocked continents at the latitude of modern-day Minnesota, he indicated. Both hothouse climates apparently maintained enough humidity to support a variety of plants around the globe. Little evidence of deserts or salt flats turns up in sediments from either the Cretaceous or Eocene hothouse. Researchers have turned up signs of seasonal dryness during the Eocene in only a few places, such as southern China. While this does not constitute proof that few deserts existed, it does suggest it. "I think that if there were really widespread dry conditions, we probably would have evidence for them," said Wing, who has led many expeditions to collect fossils from the Cretaceous, Eocene, and other warm climates. "It's not generally a very dry time, neither one of them really," Wing said. "For most places we have records, there seem to be coals and other records that suggest relatively humid conditions."

Evidence from ocean-dwelling microscopic life-forms known as forams and other indicators suggested that carbon dioxide levels during the Eocene ranged between two and six times greater than those during the interglacial warm periods of the last million years. Bursts of methane also lifted greenhouse-gas levels, at least in the early Eocene, evidence suggests (as chapter 3 covered). High levels of the greenhouse gases methane and carbon dioxide would help explain not only the warm climate but also the abundant plant growth observed. (Remember, methane soon converts into carbon dioxide and water—and does so about three times more rapidly in the warm tropics than at the poles.) Debate continues about how high carbon dioxide levels rose. But the well-documented relationship between plants and this trace atmospheric gas helped refute a line of research that suggested carbon dioxide levels during this early hothouse period might have dropped even lower than today's.

In 1999, a research team proposed that, by the middle of the toasty Eocene, carbon dioxide levels had dipped down to modern levels or maybe even as low as glacial times. To make these estimates, Paul Pearson and Martin Palmer used an experimental technique involving boron isotopes from ancient sea-dwelling forams. An inventive technique, it nevertheless required making a series of assumptions, including about the ocean's alkalinity and temperature at the time and place where the forams lived, the amount of carbon in the water from sources other than life, and atmospheric pressure. In other words, the assumptions require a bit of guesswork. Also in 1999, another group of researchers led by Mark Pagani reached similar results using a different experimental technique. Later that year, researcher Sharon Cowling considered the evidence from a land plant's point of view to reject the lowest levels reported. Her analysis showed that the minuscule values of carbon dioxide suggested would not have supported the lush plant growth recorded in the Eocene's fossil record.

In a 2000 *Nature* paper, Pearson and Palmer reported additional data. Newer samples suggested carbon dioxide levels of 2,000 parts per million or higher for the earliest part of the Eocene. The samples they had studied earlier seemed unusual in the context of other samples they collected, Pearson indicated in an e-mail exchange. Nevertheless, some of the newer data still suggested that carbon dioxide levels may have dipped down to the range of modern (interglacial) levels in the middle Eocene. The paper makes the case that at least some of these values could be reflecting cooling during this warm period—possibly even some buildup of ice in the high latitudes some 48 million through 45 million years ago. "We know there were relatively cooler periods in the Eocene, and we think we hit on one initially by chance," Pearson wrote to me, referring to the samples reported in the 1999 paper. Meanwhile, his 2000 paper with Palmer suggests that during some of the warmest parts of the Eocene, carbon dioxide levels measured about 700 to 900 parts per million—well within the range of projections for the end of this century if society doesn't work hard to reduce fossil-fuel emissions.

So the link between carbon dioxide and temperature continues to stand the test of time, albeit with the understanding that other factors also wield an influence. Cowling's efforts, which showed that glacially low carbon dioxide levels would not have supported abundant early Eocene plant growth, continue to ring true. As it turns out, an understanding of how plants function would also help enlighten how carbon dioxide levels fared in more recent times of glacial lows.

❧

By the time the Eocene gave way to the Oligocene about 35 million years ago, Antarctic ice had begun to persist long enough to make a dent in the world's water supplies. Antarctic ice sheets have generally but irregularly increased since they started building up 35 million years ago, with particular growth spurts about 13 million and 7 million years ago during the Miocene. With both carbon dioxide and temperatures dropping, the world's forests began to contract. The Smithsonian's Scott Wing noted that *Metasequoia*'s presence in North America began to wane in the Miocene, which spanned roughly 25 million to 5 million years ago. A few remnant stands survived into modern times, discovered in the mid-1940s in central China. Meanwhile, redwoods were reduced to remnant stands in California. As forests dwindled, widespread grasslands made their first major appearance, at least in the fossil record. "We know there were grasses around at least as early as the late Cretaceous," Wing said. However, "we don't have any real evidence for extensive grasslands until the early Miocene," roughly 25 million years ago.

During the late Miocene, roughly 7 million years ago, declining carbon dioxide levels likely provoked plants into evolving a new type of grass, a creative group of researchers deduced from the fossilized teeth of long-dead grazers. Jay Quade, a University of Arizona geoscientist, lead author Thure Cerling, and others used carbon isotopes in fossil teeth to distinguish what types of plants were available to extinct grazers. "From an isotopic point of view, you are what you eat," explained Quade. Because teeth are composed of minerals—mainly apatite, ironically—this dental record of consumption can last for millions of years. "Our teeth are basically rocks," as Quade reminded.

The carbon isotopes in 10-million-year-old tooth enamel from grazers speak of browsing on tree leaves and the standard grasses of the time, all of which were "C_3" plants. But the isotopes suggest a large-scale dietary shift centered on 7 million years ago to a different kind of grass, dubbed "C_4" plants. The numbers refer to how the plants collect carbon dioxide for photosynthesis. To avoid sounding like a scene from Star Wars, I will refer to C_3 plants as "standard" plants, because they came first, and C_4 plants as "low-carbon" plants, because they photosynthesize more efficiently than their counterparts when atmospheric carbon dioxide levels are low. Both persist today, depending on local climate conditions. In a sense, low-carbon grasses evolved to reproduce internally the higher carbon dioxide levels that the environment had previously provided externally. As work with modern plants demonstrates, these different approaches to processing carbon during photosynthesis, in turn, yield different isotopic signatures.

The researchers, joined by James Ehleringer and others, collected ancient teeth from every continent, using museum collections as well as a pick and shovel to unearth additional specimens. Armed with evidence of shifts in isotopic signatures occurring across many continents during the same span—between 8 million and 6 million years ago—they became convinced that the shift to low-carbon plants reflected a global change. The logical conclusion: These grasses were responding to declining carbon dioxide levels. The adaptation had another benefit. It helped these plants cling to their water supplies.

The low-carbon adaptation worked for the same reason standard plants grow better under higher carbon dioxide levels (at least within a certain range). It comes down to a case of mistaken identity. In the heat of the moment, a plant's photosynthesizing apparatus often mistakes oxygen for carbon dioxide. (Both have two oxygen atoms, so they must appear similar from certain angles.) It's not unlike the way our own blood cells can confuse carbon monoxide with oxygen. In humans, this error can lead to carbon monoxide poisoning, such as when someone sits in an idling car for too long, especially one enclosed in a garage. For plants, the consequences of mistaking oxygen for carbon dioxide are less dire. Still, the mistake costs energy. Low-carbon plants developed a way to concentrate carbon dioxide internally. With concentrated carbon dioxide feeding into the works, the low-carbon plant's photosynthesis factory wastes less energy on mistakes.

Under our modern sky, a standard plant can waste 10 or 20 percent of its sunlight-powered energy on this mistaken-identity error, known as photorespiration. High temperatures compound the problem. Low carbon dioxide levels, though, really take a toll. The lower carbon dioxide levels typical of recent ice ages can drain up to 60 percent of a plant's initial energy investment in photosynthesis—especially in places or seasons where local temperatures soar. With a doubling of atmospheric carbon dioxide, these identity errors drop by half under the same hot conditions, as Ehleringer and others describe in a 1991 paper. Given the really high carbon dioxide conditions of some past eras, these errors are negligible—even at the high temperatures that accompany the rise in carbon dioxide levels. It appears that plants invented a better carbon trap for situations featuring relatively low carbon dioxide levels along with high temperatures (at least in summer). Compared to standard plants growing under high carbon dioxide conditions, a low-carbon plant can more often close its stomata, the gates from which water escapes. So the approach conserves water as well as energy. In modern times, low-carbon plants continue to thrive in many tropical and subtropical regions where intense sunlight means they often battle high evaporation rates.

Low-carbon plants do spend some extra energy shuttling carbon dioxide into an internal "bundle sheath," the apparatus that concentrates this gas to levels roughly three times higher than our modern atmosphere. Because of this energy expenditure, the productivity of low-carbon plants drops below that of standard plants once atmospheric levels reach a certain point—about our current levels of airborne carbon dioxide.

The first evidence of bundle sheaths shows up in fossil plants about 7 million years ago, which corroborates the evidence from ancient teeth. Since then, a dozen and a half different plant families have evolved the relatively minor changes that kick-start the low-carbon photosynthetic pathway. What's more, some or all of them evolved the technique independently, experts conclude based on evidence including their location on the evolutionary tree of life. These days, about half of the 10,000 species of grasses on Earth use the low-carbon pathway. Given that it was relatively easy to make the switch—all the necessary apparatus was already there—the sudden surge in development after eons of delay suggests that carbon dioxide levels were high enough through much of Earth's history that plants didn't need to bother.

The apparent evolution and expansion of low-carbon plants suggests that carbon dioxide levels remained relatively low through the end of the Miocene. By 6.5 million years ago, the Antarctic held even more ice than it does today—although the Arctic Ocean apparently remained free of permanent ice, based on the 2001 analysis by James Zachos and others. Seas retreated, leaving high and dry the shallow Straits of Gibraltar between Africa and Spain as well as the entire Mediterranean Sea. As this epoch drew to a close, the thin cloak of carbon dioxide turned out to be too sparse to keep away the chill. When the sun cycles shifted back into a cold phase, ice began to pile up even in the Arctic. Soon after, Earth entered the Quaternary flip-flop, going from the full-blown ice ages of glacial periods to the warmer climes and higher sea levels of warm interglacials. That, too, had implications for plants.

North American glaciers finally started taking hold about 3.2 million years ago. About that same time, the Horn of Africa shrank down from rain-forest status to savanna. The Earth system began its regular cycling from ice ages to interglacials. The extent of ice cover varied during cool phases, as did sea-level rises during warm phases. But the temperatures tended to stabilize around certain thresholds. For the last 420,000 years, temperatures around Antarctica show that ice ages dipped to about 15 or 16 degrees Fahrenheit below 1950s values, while interglacials peaked about 4 or 5 degrees Fahrenheit above the modern norm, researchers led by Jean-Robert Petit concluded from Vostok cores. Comparing conditions during the last two fluctuations, then, can stand in as an example of how the planet and its plants fared during these fluctuating extremes. Our current interglacial, which remains cooler than the previous one at the moment, serves as a useful reference point somewhere in between. We're all familiar with our modern climate, moving target though it is.

The previous interglacial, the Eemian, featured sea levels encroaching roughly 20 feet higher inland than in modern times (as described in chapter 3). Like temperatures, seas rose less dramatically than they had in earlier hothouse periods, when all low-elevation polar ice melted. Still, the higher and warmer seas and hotter air worked together during the Eemian to bump up rainfall rates relative to today. Given the extra rainfall, higher temperatures, and reasonably high carbon dioxide levels, plants expanded poleward. Eastern North America took on a warm, humid climate more typical of modern-day Florida. Cape Cod's forests resembled those of North Carolina, manatees frequented the New Jersey shoreline, and osage orange trees thrived near present-day Toronto, as Robert Peters described in *Biodiversity*. In modern times, osage oranges rarely venture farther north than Chicago. The Pacific Northwest's eastern Cascades dried out relative to today. Drought-tolerant pine, oak, and juniper—species more typical of the Southwest—lived where higher humidity now supports towering mixed conifer forests, pollen research by Cathy Whitlock and Patrick Bartlein showed. Work by George Kukla and colleagues indicated that lush mixed hardwood forests thrived in Europe. And warm-water mollusks lived in the Netherlands river that gave the Eemian its name.

In England, London featured a climate currently relegated to African savanna. The unearthed bones of hippos, lions, and elephants in soil now covered by bustling Trafalgar Square capture how warmth-loving animals expanded poleward during Eemian times, as John Gribbin described in *Hothouse Earth*. Also illustrative is the northward expansion of certain beetle species. G. Russell Coope used the buried body parts of ancient beetles on British soil to estimate July temperatures based on the presence of specific beetle species. In the modern context, it would be akin to considering how climate had changed if palmetto bugs, a beetle familiar to some subtropical residents as a two-inch-long flying cockroach, expanded their presence from Miami to Boston. Researchers used the abundance of species now found only in southern Europe to estimate that July temperatures in lowland England averaged some 5 degrees Fahrenheit higher during the Eemian than today. Thus, Europe's temperature rise appeared roughly on par with Antarctica's.

While London took on the characteristics, or at least the creatures, of African savanna, much of Africa's own savanna transformed into forest. Tropical forests covered more of northern Africa than they do today and also expanded in parts of Asia, near India. Savanna expanded, too, leaving deserts to contract. The evidence also suggests that Eemian forests expanded in North America while remaining fairly stable in South America and Australia. The improvement in plant growth captured in fossils suggests an increase in global plant productivity compared to today. E. M. Zelikson and other Russian scientists analyzing Eemian peats, lake sediments, and pollen from the East European Plain concluded that forest expanded and plants grew more vigorously, enough to increase their biomass half again—that is, by 150 percent—over the amount on the plain today. Few other evidence-based comparisons of Eemian biomass exist, but there appears to have been a general expansion of northern forests where tundra now exists.

❦

The Eemian drew to a close about 118,000 years ago, as Earth's orbit and related factors led to a decline in solar power reaching the planet. Another ice age soon took hold. Northern summers grew cold again from the natural cycles in planetary orbiting. In time, ice sheets expanded their territory. To consider how glacial conditions affected plant life, researchers often look at the coldest, driest point of the last ice age, dubbed the Last Glacial Maximum, roughly 20,000 to 30,000 years ago. By then, ice had invaded most of Canada, burying its forests and grinding the remnants into the ground. The cold winds blowing off the ice sheets helped keep forests at a distance.

As the ice age advanced in North America, spruce, fir, pines, oak, and birch all lost ground to wildflowers and other herbs. Spruce forests disappeared, shrunk in area, or moved south. A favored species for Christmas trees, spruce thrives in cold climates that other species find too harsh. Spruce colonized the Midwest and the Appalachian region, where currently a variety of hardwoods thrive, as shown by pollen evidence compiled by the researchers with the Cooperative Holocene Mapping Project (COHMAP). Spruce also dropped down into the Gulf states, research by Thompson Webb and others showed. The Gulf states also supported other trees more familiar these days in chilly New England states. An analysis by the eminent pollen expert Margaret Davis found that maple didn't venture much farther east than Alabama or farther north than Missouri. Hemlock apparently eked out a living in North Carolina in a few isolated pockets of humid warmth. Evidence indicates that hemlock spread out from these remnant pockets of suitable habitat, known to ecologists as refugia, only after climate started warming again. Many other temperate hardwoods, too, took refuge in sheltered sites scattered throughout North America and Europe. In the Pacific Northwest, pollen analyses by Cathy Whitlock and Patrick Bartlein showed subalpine to alpine conditions around the eastern Cascade Range during the peak of the last ice age, with grasses and herbs inhabiting land that now hosts forests of Douglas fir and ponderosa pine.

In Europe, forests were "reduced to patches near the southern coasts," in the words of William Ruddiman, a longtime climate researcher whose 2001 book, *Earth's Climate: Past and Future*, serves as a valuable resource for students and scholars of global change. An ice sheet covered Norway, Sweden, and Scotland as well as most of England and Ireland. Only sparse vegetation grew seasonally in France, Germany, Austria, and lands north and east, most of which transformed into tundra. Even Greece was bisected by tundra. While much of Russia transformed into ice and tundra, Kazahkstan and Turkey sustained grasslands.

Tropical forests lost ground to grasses, too, as temperatures and rainfall levels both declined. The lower carbon dioxide levels also made it more challenging for trees and other woody plants to survive. Greenhouse experiments suggest that plants cannot exist once levels drop below about 100 parts per million, as Ruddiman notes. The drop is especially stressful to mountaintop trees. Much as the lower air pressure makes it difficult for mountain climbers to get enough oxygen, it makes it challenging for high-altitude plants to collect enough carbon dioxide. When carbon dioxide levels drop below 200 parts per million by volume at sea level, as they did during ice ages, they effectively drop to half that—below sustainable levels—at the height of 500-millibar winds. In our current atmosphere, 500-millibar winds (that is, those featuring half the pressure of sea-level winds) exist at roughly 18,000 feet. So do some of our tallest mountains. Declining carbon dioxide and temperature combined to push the ice-age treeline down on many tropical mountains. In the Andes, the mountaintop region inhospitable to trees extended about 3,000 feet below present-day levels, Michael Thomas and Martin Thorp noted. Although some researchers remain skeptical of the ice-age contraction of tropical rain forests, evidence from numerous pollen records supports the downward shift of tropical forests on mountains and the general conversion of many lowland rain forests to savannas.

In the Amazon, a creative approach of mapping two co-evolving species, described in *The Great Ice Age: Climate Change and Life*, by R. C. L. Wilson and colleagues, supports the concept that Amazonian forests shrunk during the last glacial. The researchers found evidence that two types of poisonous butterflies that mimic each other come in distinct pairings that differ from one region to another. This strongly suggests that ice-age rain forests split into isolated patches. The two species, *Heliconius melpomene* and *Heliconius erato*, both have bright wing colorings that act as flags to alert birds to the culinary dangers ahead. The wing designs of the two species closely resemble each other, an adaptive trait known as Müllerian mimicry. (With two poisonous species sporting the same coloring, birds learn that much faster to avoid them.) Yet the patterns shared by these two species vary from region to region. In some places, both species bear wing patterns as complex as a monarch butterfly's, whereas in other areas both species from neighboring forests may sport only a single splash of color midwing. Ten different sets of designs evolved between these species in the Amazon, with patterns varying from region to region. This provides strong, albeit circumstantial, evidence that the pairs evolved separately, presumably in remnant patches of forest—which suggests forest patches existed in moist, warm pockets within a sea of grass.

The expansion of the continental shelf area during ice ages did little to forward the cause of vegetation. Most places on Earth sported less vegetation during glacial periods than they do today. One exception was the area north of Australia, where evidence indicates that rain forests covered parts of what is now buried underwater on the continental shelf. Modern Australian rain forests occupy a narrow slice of the northwestern corner of the country, although some hold that thousands of years of land-use practices have reduced forests to a fraction of their potential extent.

Between ice sheets and the cold, dry climate, global vegetation cover shrunk down during the last ice age. In the same study of the East European Plain by Zelikson and colleagues that found a 150 percent increase in plant biomass during the Eemian, the researchers estimated that plant biomass on this stretch of eastern Europe and the former Soviet Union shrunk down to a mere 6 percent of its modern stature about 18,000 years ago during the last ice age. Plant biomass shrunk at the global scale, too, although less drastically than in the immediate vicinity of ice caps. Scientists from around the globe came together in the 1980s to compare pollen samples, lake levels, and marine plankton to estimate that ice-age plants held about 500 fewer billion tons of carbon than do today's plants, as William Ruddiman, a leading member of the research team, explained in his 2001 book, *Earth's Climate: Past and Future*. That's about seventy-five years of fossil-fuel emissions at modern rates. Later studies reached slightly different conclusions, ranging from no major change to a halving or greater decrease of biomass, with most estimates suggesting a decrease in ice-age biomass of 300 billion to 1,000 billion tons compared to today. At any rate, no estimates proposed that ice-age forests and grasslands held more biomass. All the evidence points to a decrease in plant biomass during ice ages.

❧

If we could put the globe on a scale that measured weight gain solely from plant growth, we'd find that it increases overall during the warm, wet periods of the past 100 million years. It's all those carbohydrates produced during the longer growing season. Forests generally flourish during warm periods and shrink in stature and range during ice ages. As ice sheets expand across the landscape, they bury forests and the carbon within them, sealing them off from the atmosphere. Similarly, the deep ocean apparently manages to cling to its carbon supplies during glacial times, locking up carbon far from the surface and thus preventing an exchange with the air.

Some of the carbon buried under ice sheets joins the atmosphere once again when glaciers retreat during interglacial warm periods. The newly exposed area contains rich soils the glacier plowed up from farther north and pushed before it. And, of course, the mix contains some excellent compost derived from the long-dormant forests and grasslands buried by the glacier's descent. The fertile midwestern soil that farmers love came from retreating glaciers. Before agriculture and development took over in modern times, the ground fertilized by glacial till would have welcomed incoming vegetation expanding during warm periods. As glaciers melt, they release water to wetlands and coastal waters and relinquish ground to forests. Spruce trees, one of the most cold tolerant of their woody brethren, drift in from the southern boundaries of glaciers, awaiting the promising moments when ice becomes water. Other plants are poised to replace them as they move north. Warmth-loving trees and plants expand from their ice-age refuges to repopulate the landscape.

A warming climate and the higher precipitation rates and carbon dioxide levels that come with it globally boost plant growth. Anywhere sufficient moisture combines with warming temperatures, life expands and pulls down carbon. This can be seen as a Gaian response in the classic sense, in that life is involved. Meanwhile, the tendency for precipitation to increase during warmer times serves as a Gaian physical response, one that fits into Lovelock's 1979 expanded concept of Gaia as a coupled system involving life and its physical environment. This also fits with the declaration signed by a thousand scientific delegates at a 2001 international scientific conference in Amsterdam that "the Earth System behaves as a single, self-regulating system comprised of physical, chemical, biological and human components."

To look at the nonhuman influences that operate across millennia or more, we have to look deeper. If we look beneath the surfaces of forests and wetlands and the sea, we will find some of the carbon these systems collected from the atmosphere over the years. Globally, soil holds several times the amount of carbon found in the world's plants. Soil serves as a long-term medium for storing carbon, especially where the ground is moist. Where it's really moist, wetlands act as a soggy sink for carbon. Their coal-black soils hint at the density of this element. In fact, wetlands launch the lengthy process of coal formation. Thus, as the next chapter explains, wetlands and forests also help balance carbon dioxide levels at time scales beyond the centuries and millennia that it takes for plant growth to stabilize after major climate changes.

6

An Internal Cleanse

The Additive Power of Soils and Wetlands

The word "swamp" often conjures up dark images, at least in Hollywood. If the movie you're watching starts showing images of buttressed trees knee deep in water with moss draped over their branches, a scary scene probably lurks around the bend. For portraying gloom and doom, the forested wetlands known as swamps have a reputation on par with stormy nights. Yet both rainstorms and swamps share life-giving water with the landscape. Sure, some pests will take advantage of the moisture—mosquitoes, even leeches. Maybe the occasional presence of these bloodsuckers helps explain the silver screen leap to werewolves and vampires. Yet this focus on this one pesky feature of some wetlands ignores their positive role in so many things that promote and improve life, including for humans. Besides, it's not as if draining swamps will get rid of mosquitoes. As long as the local landscape or landfill provides puddles the size of pop-tops for a few days at a time, mosquitoes will thrive. Wetlands, on the other hand, need abundant water for at least season-sized chunks of the year to carry out their many services to the planet and its life-forms.

Wetlands have a special talent for collecting carbon that relates to their remarkable productivity. From Puerto Rican mangroves to Louisiana bayous to Alaskan peatlands, and all sodden pockets in between, wetlands offer a productivity that is almost unparalleled. This productivity is especially potent for drawing down carbon when combined with slow decay rates. Decay stalls in stagnant waters. What's more, wetlands' low-lying position on the landscape allows them to collect organic material from distant forests and grasslands, further overwhelming the local biota's ability to decompose it all. Perhaps it's this hint of incomplete decay that earns swamps a dark corner in the human psyche. But it also gives them an important Gaian role in regulating planetary greenhouse gases. Chapter 8 will touch upon some other important wetland functions, including reducing flood damage, moderating temperature and water cycles, and purifying water. This chapter will focus on the role of wetlands in building land and taking up carbon dioxide, in the context of their emissions of another greenhouse gas, methane. It will also touch upon the carbon-collecting role of soil in general.

Soils, too, can hold a scary place in our subconscious. Perhaps because we bury our own six feet under, our impression of soils also suffers from an association with decay. Landfills don't help their image. In reality, the decay processes occurring in soils and wetlands make life all the more sweet. Decomposition of dead organic matter keeps soils productive, churning over nutrients for the next generation. Spiders and mites and other creepy crawlers keep the fertility factory humming. Decay has a bad rep, but it's an essential part of the system. It's what keeps the world going round. Without it, all the nutrients and carbon would have been locked up long ago. The drop in carbon dioxide that ushered in the lengthy ice age of the late Carboniferous and Permian suggests what can happen without decay to keep carbon cycling. A popular theory holds that this ice age 300 million years ago traces back to the evolution of wood, and its preservation in wetlands, before the evolution of the critters that could decompose it back into usable nutrients.

The planet's wetlands and soils work with the plants, the atmosphere, and the changing climate to balance the carbon cycle. They soak up some of the excess carbon dioxide in times of high productivity. And they can dry out and fling carbon-laden dust to the wind during cold, dry times that come with glaciers. Methane releases in freshwater wetlands can counteract some of the cooling effects from the drawdown of its fellow greenhouse gas, carbon dioxide. This seems to be less of an issue in shallow wetlands, as William Mitsch explains in the 2007 book, *Wetlands*, that he co-authored with James Gosselink. But there's no denying that peat formation is the first step to coal creation. And coal pulls down carbon for the long haul—or at least until some enterprising biped figures out how to extract and burn it. What's more, some soils can actually consume methane. Clearly, the substrate beneath our feet, whether terra firma or the soggy soils of wetlands, helps regulate the planetary thermostat. You would think this crucial carbon-collecting service would earn wetlands and soil a pedestal in our species' collective unconscious instead of a subconscious fear of them.

❧

As a society, we treat soil like dirt. To most of us, soil is just something to stomp around on, a substrate to keep our houses settled and our streets in line. Gardeners know otherwise. By collecting potato peels, coffee grounds, and other plant remnants in a compost heap, gardeners of every ilk recognize a basic truth about soil that sometimes is overlooked by global-scale climate models: Some plant organic matter transforms into soil organic matter even after the rest has decayed back into carbon dioxide. Gardeners and ecologists call this humus. "Humus, by the way, is when you can't tell it's a banana peel," as Lynn Margulis put it during a talk at a 2006 conference on Gaia theory held in Arlington, Virginia. The value of humus has become obvious as more gardeners and farmers turn to organic farming to make a healthy living or meal. They recognize the natural fertilizers, such as nitrogen and phosphorus, that mingle with the carbon in humus. The nutrients and organic carbon mainly comprise material shifted from plant to soil. Yet even now, many global-scale climate-vegetation models base their carbon dioxide calculations on the assumption that fallen trees decay instantly. In fact, decay can take decades or centuries. Even more important, an undefined amount of carbon from logs, branches, and leaves transforms into long-lasting soil organic matter. Again, forests and wetlands excel at converting plant material into soil carbon and peatlands, repectively.

In the humid tropics, the carbon moving into soil can amount to a drawdown of about 1,600 pounds of carbon dioxide per acre of soil per year. That's what researchers Ariel Lugo and colleagues reported in a 1986 paper when comparing the amount of carbon in land under farms to land that had been left fallow to form pasture or convert back into forest. That, of course, is carbon dioxide beyond the amount taken up by the plants themselves. These additions bump up carbon numbers under any plant-covered soil that doesn't face the farmer's aerating plow. A 2000 review of several studies led by William Schlesinger of Duke University found comparable rates of increase under temperate broadleaf forests, with grasslands and conifer forests generally accumulating carbon about half as quickly.

Warm, humid climates generally sprout the biggest plants, as described in the previous chapter. Meanwhile, bigger plants tend to leave a deeper trail of carbon in the soil. So perhaps it's no surprise that the warmest, wettest environments generally contain the deepest soils with the most carbon. Tropical forests store more carbon in the top 10 feet of their soils than do other ecosystems, as Estaban Jobbágy and Robert

Jackson found using thousands of soil profiles collected over the years. (Wetlands were not included as distinct entities in their analysis, though. They're in a class of their own, as the next section will illustrate.) By their assessment, the carbon-rich tropical soils hold twice the amount of carbon as those under boreal forests and tundra put together. Even temperate forests, such as the Pacific Northwest rain forests and Northeast deciduous forests, hold nearly as much carbon as the boreal forest and tundra combined. Savanna systems and other tropical grasslands, meanwhile, store twice the amount of carbon as temperate grasslands. Any vegetation-covered soil is collecting carbon, with the lushness of the plants a reasonably good indicator of the vastness of the carbon supply beneath the surface.

All in all, the world's soils hold about three times the amount of carbon found in the vegetation growing in them, Jobbágy and Jackson found. The researchers assessed roughly the top 10 feet of soil, whereas most studies focus on only the top few feet. This depth probably captures the bulk of many high-latitude systems. The depth of northern peats, for instance, averages about 7.5 feet deep. Tropical soils, though, can extend much deeper. Despite the steep slopes in Puerto Rico's forests, the soils there typically extend 20 feet or more below the surface. "In Bisley, weathered clays get down to six or nine meters, and they're still nutrient rich," explained Frederick Scatena, a geoscientist who spent more than a decade leading research projects in the Bisley section of Puerto Rico's Luquillo Experimental Forest. That's roughly 18 to 28 feet. "All of those have the capacity to take in more organic matter," he added during our conversation. The amount of carbon in the world's soils is increasing where trees and shrubs are "invading" grasslands, a global phenomenon described in the previous chapter.

Steve Archer, a professor in the University of Arizona's School of Natural Resources and the Environment, examined the changes occurring after mesquite trees invaded Texas grasslands. "We've got basically seven- to eightfold more carbon in the system now than we would in a pristine grassland," he said, noting that this included the changes in plant biomass as well as soil carbon in the most active root zone. Soil carbon also was rising in other parts of the country where trees and shrubs were invading grasslands, as Archer and his colleagues, including lead author Alan Knapp, found when comparing the results of similar studies at eight different sites. Because the researchers had compared different soil depths at different sites, they couldn't directly compare soil carbon changes among all the sites. Using six of the most comparable sites, though, they estimated that soil organic carbon had increased by about 7 percent since the arrival of woody plants.

❧

Trees and shrubs speed up the transfer of carbon from the air to the soil for several reasons. Wood is more dense and less nutritious than leaves. Leaves nourish a variety of creatures, including soil animals that can quickly devour fallen leaves, typically within a year. But few creatures besides termites find wood palatable. Leaves can only grow so thick before they cease to function as photosynthetic factories. Their finer form, then, makes it easy to break them up into smaller, more digestible pieces. Waves and winds can smash them to bits. Even a child can crush a leaf. Splitting wood, on the other hand, often requires a concerted effort with a saw or an axe. Wind can split a tree, but the log it leaves behind often endures for decades, centuries, or longer. All plants release carbon directly into the soil through their roots, sometimes to curry favor with microorganisms that can provide nitrogen or phosphorus in exchange. Trees, though, tend to have the deepest and most abundant root systems. Their roots can reach dozens of feet down in the soil, exuding carbon while alive and potentially creating pockets of carbon upon death. What's more, the quantity of large, woody roots continues to increase throughout the life of a tree. This suggests that old-growth forests could continue to accumulate carbon beneath the surface for many hundreds or even thousands of years, depending on the longevity of its trees—and its deadwood.

The tangle of deadwood marking Hurricane Hugo's path turned Puerto Rican forests into obstacle courses for a few years after the storm, and many partially decayed remnants remained visible during my 1998 forest survey after Hurricane Georges. By 2006, most of that mess had decayed, a testament to the generally high decomposition rates in the moist tropics, Ariel Lugo mentioned in a telephone conversation. Still, he noted that many large logs remained virtually intact on the landscape—and appeared to have the potential to endure for many decades to come. One or two big trees typically weigh in with most of a tropical rain forest plot's biomass, as shown by Lugo's research with lead author Sandra Brown. Large or small, about half of a tree's biomass comes from carbon. With the rare toppling of one of these "big boys," as Lugo fondly calls them, the carbon they contain can remain immobilized for at least decades and potentially centuries.

I had the chance to meet a few of these enduring pieces of deadwood in 1989 during a brief stint working on a project set up in the Caribbean National Forest by Lugo and Marinelly Vellilo. I recall one gray-barked giant with a tag indicating that it was last seen alive sometime before 1966. When I rested my back against it and flung my arms out, the still-intact trunk extended about elbow to elbow. It stretched out along the forest floor for the length of a three-story building. These big trees don't topple easily—but when they do, they leave their mark on the landscape.

Certain species are notoriously resistant to decay, as forest researchers and anyone who works with wood knows. "When you start looking at the hardwoods, like mahogany and asubo, those remain on the forest floor for a long, long time," Lugo observed. Wooden beams from asubo continue to provide structural support in Old San Juan's colonial buildings, some of which go back to the seventeenth century. The wood remained intact in the twenty-first century, as I learned from an old friend known as Baba, a local artist who carved beautiful sculptures out of ancient asubo beams discarded by renovators. A cousin to asubo, the sapodilla—also known in English as the chicle tree, which inspired the name for Chiclets gum—would take hundreds of years to decay even on the ground of a tropical rain forest, judging from the resistance of foot-wide logs in tropical Mexico to the first four years of decay. That research, by forestry professor Mark Harmon and others, indicates that larger logs would take even longer to decay. Certain tropical woods also contain protective secondary compounds, such as the white latex that will ooze from the chicle tree's slashed bark. Think of them as embalming fluid for trees. These compounds also help redwood trees resist pathogens for the thousands of years that they can live—and help keep wood intact after the tree's death as well. This longevity makes redwood a favorite among carpenters and contractors. Secondary compounds allow many moist forests to keep deadwood on the ground for decades.

For similar reasons, a walk through some of the old-growth rain forests of the Pacific Northwest will require many forays over tree trunks, unless people carrying chainsaws have forged trails through them. The same holds for many forests around the world, especially those impacted by hurricanes or other disturbances. Look around the next time you're in a North American forest where people rarely harvest fallen trees for fuelwood. You might be surprised at the sheer volume of deadwood littering the forest floor. That's how my colleagues and I felt, anyway, once we actively began looking for fallen trees during a study on uprooting dynamics. Researcher Connie Woodhouse had a similar experience when she returned to some of her Colorado research sites seeking dead trees to push a tree-ring timeline further back into the past. As she noted in passing to a group of us attending her 2008 talk at the University of Arizona, "Most sites, there's all kinds of deadwood underneath that you don't even notice unless you're scavenging for firewood or something."

In theory, old-growth forests at some point would reach a "carbon steady state," with the amount of carbon dioxide decaying from the logs on par with the amount of carbon dioxide being consumed by the living forest. This kind of thinking drives the push in some small circles to replace old-growth forests with "vigorous" young plantations. Yet storms, bug attacks, fire, and other disturbance events interfere with a forest's attempts to reach this theoretical steady state. In modern times, changing carbon dioxide levels, temperatures, and precipitation rates all keep forests from settling into anything near a steady state. The changing atmosphere and climate has actually been promoting growth in many of the world's forests, old and young alike (as the last chapter explained). Increasingly, the concept that modern old-growth forests release as much carbon as they consume is seen as a misconception. Still, it remains challenging to pinpoint how much carbon is finding a long-term home in plants and soils.

In the later stages of decay, logs can be almost indistinguishable in color from the underlying soil. At a site in Colorado's Sangre de Cristo Mountains where I did some of my dissertation research, the most decayed logs would remain as mere outlines on the surface, distinguishable in relation to the ancient pit formed in their uprooting but otherwise blending into the landscape. Even their texture becomes crumbly. A good kick or a canine's enthusiastic digging can turn a long-dead piece of wood into a pile of mulch that even feels like soil. At some point, organic matter becomes soil organic matter.

It's difficult to draw a boundary on this process. The transitions can blur around the edges. But in a study in pristine Alaskan spruce-hemlock forests, Bernard Bormann and colleagues measured consistently increasing carbon stores in soil mounds created by trees uprooted during windstorms over the past several hundred years. They were able to detect wood from tree trunks in mounds created by a 1930 windstorm. In the mounds from an 1830 event, however, the wood had morphed into a strip of carbon-rich humus. Organic matter sinking deeper into the ground moves further from the clutches of oxygen, with its propensity to convert carbon back into carbon dioxide. Thus the carbon from this organic matter becomes more "recalcitrant"—that is, more resistant to the lure of an airborne existence with oxygen. Even in waterlogged soils with slow decay rates, most of the long-lasting soil organic matter started out as wood, whether in roots or logs.

❧

Scientists are well aware of the potential for the world's soils to help balance the excess of airborne carbon dioxide. Policy specialists have even proposed that agricultural land could take up carbon dioxide if farmers adopted "no-till" planting approaches. Tilling involves breaking and churning the soil, thus exposing its carbon to ravenous oxygen in the air. This plowing makes the soil more susceptible to erosion, too, another potential pathway for putting carbon back into circulation. So laying off the ground-moving machinery can allow even cultivated soil to retain carbon. Meanwhile, letting natural systems recover from previous disturbance, whether caused by intentional land clearing or natural processes, also adds carbon to the soil.

On the other hand, some worry about whether soils can retain their carbon as the temperature heats up. The northern soils currently left frozen much of the year are a case in point. Warmer temperatures tend to increase the release of carbon dioxide in respiration. Yes, even trees and soils respire. Like animals, plants release carbon dioxide when they use some of their carbohydrates to provide energy. Soils, too, respire, thanks to energy conversions from the minute animals in their midst. This has led to valid concerns that warming soils might release massive amounts of carbon dioxide. Yet an experiment in spruce forest in Sweden suggested it may be premature to assume the extra heat will lead to carbon releases. Research by Paul Jarvis and Sune Linder turned up no detectable difference in the amount of carbon dioxide coming from soils kept 9 degrees Fahrenheit warmer than neighboring soils during a five-year experiment. The researchers found that warming the soil a few inches below the surface increased carbon dioxide output by only 10 percent, within the normal range, while simultaneously boosting tree growth by more than 50 percent. Other studies also are challenging the assumption that a rise in temperature automatically means a rise in the amount of carbon dioxide released by soils via decomposition. In a *Nature* paper comparing studies from eighty-two sites on five continents, Christian Giardina and Michael Ryan found forest soil decomposition rates "remarkably constant" around the world. While the decay rate of the leaf layer on the forest floor does speed up in hot and humid climates, the mineral soil below—where most soil carbon resides—is more insulated from temperature effects than standard computer models assume, they concluded.

Many factors will affect the amount of carbon making its way into soil stores—including some of the same factors that are improving forest growth, such as carbon dioxide fertilization, longer growing seasons, and warmer temperatures at high altitudes and latitudes. The world's vegetation tends to blossom productively under warm climates when left on its own (as the previous chapter described). Given the strong link between plant productivity and underlying soil carbon, it seems logical to expect that soil carbon will increase where vegetation is becoming more productive.

Soils have another talent that makes them important to greenhouse-gas discussions. While the plants growing above them consume carbon dioxide, and in time transfer some of it beneath the surface, the soils themselves often consume methane. Or rather, the bacteria they support do so. (Once again, as mentioned in chapters 1 and 2, a climate-control skill comes down to bacteria at the root.) Mohammad Aslam Khan Khalil, editor of the 2000 book *Atmospheric Methane*, and colleagues estimated that soils take up about a quarter of the methane emitted globally by their more sodden brethren, wetlands.

Some of the key work showing that soils take up methane involved Michael Keller, a scientist with the Large-Scale Biosphere-Atmosphere Experiment in Amazonia. His study with colleagues found that the soils under tropical rain forests in Panama were taking up about four times more methane than those under cleared areas with cattle grazing on them. The hotter temperatures and drier conditions that come with the removal of forest cover apparently slowed down the activity of the bacteria that dine on methane. Trampling by cattle compounded the problem in the Amazon as well as Panama, Keller explained during a 2006 talk at the University of Arizona about his ongoing work in the Amazon. "They also pound the ground here—they have small, little hoofs, so they compact the soil," he said. Selective logging, too, decreased the soil's ability to absorb methane, as Keller and colleagues found in their follow-up study in the Amazon. In fact, many of the soils under logged forest actually started emitting methane. "The message here is methane is biological," Keller emphasized.

Scientists know methane emissions come from systems featuring low oxygen levels, such as the gut of cattle and waterlogged rice paddies, landfills, and wetlands. It's fairly straightforward to measure how much of it resides in the air. But, as Keller noted, the actual amount of methane being released from or absorbed by specific systems remains mysterious. The range is large enough that a 2006 report by Frank Keppler and colleagues suggesting plants themselves were releasing methane theoretically could have fit into global estimates—it would have just required shifting some of the blame from wetlands to forests. As it happens, the 2006 report has not held up to further scrutiny, although it continues to inspire debate. Keller expressed frustration at the wide range in estimates for such an important gas. "Here we are in a world with a rapidly changing atmosphere, and we can't tell anybody where we're at," he said.

The comings and goings of both methane and carbon dioxide in soils increase phenomenally in soggy soils. Impressive as it is, the carbon-storing capacity of soils pales in comparison to what wetlands can do. Swamps and marshes boasted belowground carbon stores about four times *higher* than that found under tropical forests, even though their aboveground biomass averaged about two to three times *lower* than tropical forests, as Wim Sombroek and colleagues found. Other studies support their analysis. In Virginia, Chesapeake Bay floodplains supporting red maple, green ash, tupelo, oak, and swamp cypress pulled down about three times more carbon than a typical tropical forest, based on research by Gregory Noe and Cliff Hupp. In Thailand, drawdown into peat swamps reached ten times higher, with some 16,000 pounds of carbon dioxide per acre of land annually coming down to earth.

❦

Where Puerto Rico's Mameyes River meets the Atlantic Ocean, the turquoise waters spread out into a knee-deep estuary. Its source, the Luquillo Mountains, looms over the stream from a couple of miles away, cloaked in hundreds of different tropical tree species. From this distance, the trees of many shapes and sizes blend into one green garment, visibly textured, undulating with the mountain slopes. On this summer day, clouds shroud the highest peaks—as usual. The water released by previous clouds pummeled the canopy, crashed over boulders, and slid across clay. Hours or months later, depending on its path, the water that once fell as rain passed through wetlands fringing the river, where crowded palm stands or buttressed mangrove trees slowed its passage and filtered out sediments. Finally, it settled into this calm expanse where *agua dulce*, a sweet-laden endearment for freshwater, mingles with the salty sea. This unusual blend creates an estuary, so similar in sound and service to a sanctuary. Estuaries harbor a variety of species, serving many marine creatures as a nourishing nursery during the earliest and most vulnerable stage of their lives. Also providing shelters and nutrients are mangrove trees propped up on spindly roots, tolerating water and salt levels that would slay most other species. Estuaries and adjacent wetlands shelter shrimp, crabs, and other delicacies that decorate dinner plates across the island.

The people who live around bayous, mangroves, and marshes have long appreciated the bounty of wetlands, including shrimp, crab, lobster, and a variety of fish. On the Louisiana bayou, even the poorest person can collect the main ingredients of crab jambalaya, shrimp creole, or blackened catfish, reminded Windell Curole, a lifelong resident of southern Louisiana. "I hear people talking about thinking outside the box. In this place, we don't make boxes. The way we are is because of the land. The reason people have a good life and enjoy themselves is if you've got a full stomach, everything else is extra," said Curole, noting that he'd had fresh oysters the previous day. Shrimp is another favorite delicacy spawned in estuaries. Mike Tidwell described the local shrimp season in his poignant 2003 book, *Bayou Farewell*, which beautifully captures the celebratory Cajun culture and the ongoing threats to the environment that sustains it:

> Cajun fishermen like to say that when God created shrimp, he created food for just about everybody. Anything in the marsh that can swim, crawl, walk, or fly is, in the spring, most likely looking for shrimp. But it's the fish who go especially wild. [Speckled trout, drumfish, and southern flounder] . . . each of these fish species can eat a full third of its body weight in shrimp every day. But despite all the assaults, the shrimp population remains miraculously huge, almost beyond comprehension. In early May, across vast stretches of interior wetlands, the water is literally boiling with shrimp when the oldest of the autumn-spawned class reach a length of around four inches and begin the reverse migration back to the sea, their goal to live and reproduce out in the Gulf, completing the cycle.

That's when Louisiana fishermen join the feast. Because of a population-protecting ban that usually restricts them from shrimping within the bayou itself during the start of the early May migration, they must wait at the mouth of the estuary for the nighttime mass exodus of the favored brown shrimp to seek their share for jambalaya, gumbo, and general livelihood.

The leaves and other organic matter arriving from throughout the landscape keep the system running in high gear, nourishing shrimp and a variety of other creatures. Shrimp will eat everything from decaying plants to algae, from other crustaceans to fecal pellets. But leaves put bling into the food chain. Ecologist Ariel Lugo recalled a recent trip to Micronesia, where he described their importance to some of the local fishermen. "The raw comment that a mangrove leaf is what feeds the fish makes the local people's eyes pop out," Lugo told me. "That's a surprising thing because you never see the fish eating a leaf." But the leaf nourishes the creatures that fish eat, including bacteria and fungi—and, of course, shrimp. Many shrimp species are filter feeders, collecting floating carbohydrates, including tiny remnants of decaying leaves. Mangrove systems are so productive naturally that converting them into managed shrimp farms actually tends to decrease their productivity, noted Robert Twilley, a Louisiana State University professor whose many research projects involving wetlands have encompassed this topic.

The productivity found in mangroves and other wetlands comes down to earth in wetland soils. While the other soils of the world collectively store about 2,300 billion tons of carbon, the tiny share of the land that represents wetlands alone holds between 200 and 800 billion tons, Nicolas Gruber and colleagues reported in a 2004 assessment. A middle value of 500 billion tons amounts to the removal of about 1.8 trillion tons of carbon dioxide, most of it since the end of the last ice age. As the broad range in the estimate indicates, just how much they hold remains uncertain. It's not just the amount of soil carbon that's at issue. Estimates of the area of wetlands on the planet range from about 2 to 4 percent, usually settling in the middle, although William Mitsch and James Gosselink set the figure at 5 to 8 percent in their 2007 book *Wetlands*. While every ecosystem area measurement leaves some room for waffling, wetlands pose special challenges because they often exist within other ecosystems. They're the soggy meadows and river-hugging forests that you might not recognize as wetlands until after you've buried your foot in the mud.

"We have a tendency to think only in terms of the large wetlands, the Okefenokee and Everglades. But in general, wetlands are small portions of landscapes," Lugo said. "When you look at the landscape, you always find that they're dotted with little pockets of wetlands. You take all those little ones and add them up, and then you have the equivalent of an Everglades." About two-thirds of the world's wetlands support trees, so they sometimes blend into the forest. Even the bigger wetlands are scattered across the continents, as shown in a map included in a 1990 book, *Forested Wetlands*, that Lugo co-edited with colleagues Mark Brinson and Sandra Brown. The book, part of an Elsevier series on *Ecosystems of the World*, remains one of the few global syntheses out there on these crucial but often overlooked systems. Researchers still were ignoring wetlands in most carbon-cycle models, at least as of the 2004 assessment by Gruber and colleagues.

It doesn't help that there's no international consensus on what constitutes a wetland. U.S. definitions, meanwhile, vary with political whim. In the scientific realm, one common definition features their value to birds, while another popular one gives options including the recurring saturation of soils and the distribution of plants adapted to inundation. Fuzzy definitions notwithstanding, ecologists can recognize a wetland when they see it. Wetlands include shallow lakes, river floodplains, and tidal zones as well as any lands that are waterlogged or at least soggy for some part of the growing season. A key point not usually noted in definitions: Wetland soils and waters often run dark with carbon, as all wetlands pull carbon beneath the surface. Mangroves and other systems featuring water flowing through them, however slowly, tend to export some of their carbon to the sea. In less aquatic systems, such as peatlands, carbon may build up in underlying soils or sediments. Peats are just another type of soil, a family known as histosols. It's really just a matter of degree—once soil carbon derived from plants comprises about half of its mass, the soil is considered peat.

❧

Where water pools, carbon sinks into the earth. Where water flows, swamps and mangroves thrive. Where water lingers near the surface, peatlands can flourish. Peatlands often feel a bit spongier underfoot, at least when dry. So, if we think of forests and wetlands as carbon sinks, we can think of peatlands as carbon sponges. As with wetlands, if you wring them out to dry, some of the carbon they've taken down will escape back into the atmosphere. Peat offers a relatively easy way of measuring the amount of carbon drawn down into certain wetlands. The decaying plant parts actually create land—either directly below the living plants or downstream at the edges of deltas. Branches, roots, and other woody parts have the best chance of escaping decay. Like trees, peat is typically composed of about half carbon by dry weight. Also like trees, peatlands add new layers at the surface. And while trees can transform into charcoal upon burning, peat can transform into coal with heat and pressure. As peat surface layers build up, the pressure and heat bearing down on the belowground peat also build. In time, peat can harden into coal, maintaining most of its stored carbon but in more compact chunks. It takes about 10 feet of peat to form a foot of the hardest coal, anthracite, based on estimates made for the Everglades and the Okefenokee swamps.

Bogs, fens, mires, mangroves, heath, fen meadows, swamps—all of these can form peat under the right conditions. Most of us have come across them at some point, perhaps in hardwood forests around the Great Lakes or in cedar swamps of the Southeast. Peatlands favor wet regions where rainfall during the growing season occurs in regular doses rather than in seasonal bursts. At the global scale, this means they are partial to the regions where large-scale circulation patterns (described in chapter 4) yield moisture regularly. Most of the world's active peatlands thrive either around the equator or in the north where landmasses abound at their favorite latitudes, a swath that centers on 55 degrees. These northern peats began forming within a few thousand years of the retreat of the ice sheets. The former Soviet Union has the biggest cache, followed by Canada. Other lands featuring soggy states include the United States, the United Kingdom, and several icy lands such as Finland and Sweden. Another 10 to 20 percent thrive in the tropics, hovering around equatorial lands in Indonesia and Malaysia. The rest are scattered around the planet in places where moisture can keep decay at bay long enough for discarded plant material to stack up.

Productivity in peatlands follows a trend similar to ecosystems in general, with higher productivity rates under higher temperatures when moisture suffices. Given stagnant water near the surface, peat production rates in sphagnum moss systems from around the world generally increased with temperature, Urban Gunnarsson found. In the moist tropics of Indonesia, peatlands can add about 6 inches of vertical growth a century. That's about four times faster than the typical growth rate of northern sphagnum moss. According to *The Cretaceous World* by Peter Skelton and colleagues, coals formed about three to six times faster in the tropics than in cooler climes during this hothouse environment. In general, then, warmer environments yield the most coal—as long as adequate moisture exists.

Still, year-to-year variability and other factors can influence the annual tallies from peatland systems. Higher temperature penetrating the surface increased decay rates of underlying peat, in a study of Canadian black spruce forest mixed with mosses by M. L. Goulden and colleagues. The researchers measured subtle variations in local carbon dioxide levels to conclude that late-summer thaws reaching about a foot and a half below the surface pushed the system into releasing slightly more carbon than it took up over the year. On the other hand, a nearby wetland area that was more waterlogged than their spruce site continued to take up carbon during these same years, the authors noted.

The height of the water table has a bigger effect on peatland health than temperature, many researchers agree. When left on its own, peat growth tends to reflect the ups and downs of the water table, with episodic growth spurts and slowdowns, as researchers Lisa Belyea and Nils Malmer found in studying a Swedish mire. A rising water table tends to spur on peat growth. Examining mire cores stretching back 5,000 years, they found evidence indicating that rising water tables would engulf more litter, thus reducing decay rates so that peatlands could continue to build up. Falling water tables led to a slowdown of plant production, thus exposing more organic matter to decay. Peat growth under Puerto Rican mangroves similarly kept pace with sea-level rise—which, in this case, reflected sinking of the land—over the past 4,000 years, Lugo told me, indicating that he was referring to work by Professor Elvira Cuevas of the University of Puerto Rico that had not yet been published as of early 2009. But peatlands often were able to maintain themselves through local climatic changes. Every once in a while a system can change abruptly in response to a relatively small change in moisture, Belyea and Malmer found. Apparently, thresholds exist that can tip the balance. The shift can bring a change in species or a shutdown of peatland construction. In general, though, regular rainfall during the growing season can support peatlands under many conditions, including after they've raised themselves above the local water table. The end result: Peatlands seem to have an ability to maintain themselves, to some degree, as the water table rises and falls.

Such local responses can translate collectively into a Gaian function at the global scale. In times when sea-level rise pushes up local water tables, peatlands can grow faster. As the water table lowers, such as when glaciers start collecting the world's water into frozen towers of ice, even peatlands that escape burial will begin to degrade. In the process, they release carbon dioxide into the atmosphere. The decline of peatlands that comes with the decay of carbon stores can be its saving grace. As the peat surface drops locally, it may eventually meet back up with the water table. Meanwhile, if many peatlands are registering a similar natural response to declining water tables, the global effect of their decay may be to slow down the cooling that is stealing away the water. The opposite situation can occur in times of warming. Rising water in the sea and the connected groundwater table will boost the productivity rate of existing peatlands, potentially allowing them to stay on top of the situation. Rising water tables will also encourage the expansion of new peatlands, as more of the land becomes waterlogged. Overall, this extra peat production will pull some carbon down from the atmosphere, thus potentially moderating the warming. The methane issue adds an extra twist in this situation.

Peatlands' hearty appetite for carbon sometimes leaves them with a metaphorical case of indigestion. The result: release of natural gas. The powerful heat-trapping property of methane, the same natural gas that heats many U.S. homes in winter, undoes some of the cooling the wetlands provide by taking up carbon dioxide. A 2004 paper by Eric Davidson and Paulo Artaxo suggested that methane emissions from the Amazon—where a million square kilometers (roughly 400,000 square miles) of forest floods during a typical rainy season—counteract about a third of the region's reduction in global-warming potential via carbon dioxide uptake.

Before getting into that, though, an important thing about methane deserves highlighting. This natural gas typically survives only a decade or two in our modern atmosphere. What happens to it then? It transforms into carbon dioxide—and water. Like methane, both are greenhouse gases. Yet water clearly has roles that go far beyond its heat-trapping abilities. Water is essential to life. And it's involved in evaporative cooling and in forming sun-shielding clouds. In fact, drought is a bigger threat than rising temperature to most ecological systems and croplands. So the role of wetlands in keeping water circulating should not be ignored when considering their impact on planetary climate. From a Gaian perspective, it's interesting that water-loving wetlands have a means for moistening the air. Almost all of the conversion occurs in the layer of the atmosphere where weather functions. What's more, the physical conversion of methane to water speeds up under warmer conditions. Once again, we see that ecological systems somehow work with the laws of physics to create conditions that sustain life.

❦

Peatlands are taking up carbon dioxide as they grow and releasing some of it if they decline. But then there's methane. Part of the stew known colloquially as swamp gas, methane naturally arises as decay or fermentation occurs in situations where oxygen is lacking. In landfills, the muted decay of old hot dogs, yesterday's newspapers, and other organic matter yields methane that its operators can sometimes tap. If they're not actively collecting this natural gas to sell as fuel, they're probably burning it off in flames fringing the dump's perimeter. The ability of cows to produce methane in their guts has fueled many a joke. We humans, too, have this gas-producing ability (much to the glee of many a four-year-old), although not anywhere on the scale of grazers. Emissions from cows, rice paddies, and wetlands add up to the biggest sources of methane globally. Other sources that add up include landfills, sewage, pipeline leaks, and termites. But wetlands can hardly shoulder the blame for recent rises in methane levels. After all, the area of land they cover has shrunk, thanks to farming and development, even as methane levels have skyrocketed.

Since the last ice age, northern peatlands have cooled down the globe more than warmed it up, despite the extra punch of methane as a greenhouse gas, a 2007 analysis by Steve Frolking and Nigel Roulet found. Although methane has more warming power than carbon dioxide, it resides in the air only for about a decade to a quarter of a century. Meanwhile, the carbon dioxide swept from the air into deep storage in peatlands generally remains locked up for centuries or more. Another important factor involves the differing amounts of the two greenhouse gases. While methane releases from northern peatlands are measured in *millions* of metric tons, carbon uptake is measured in *billions* of metric tons. After weighing these and other factors, Frolking and Roulet concluded that northern peatlands alone had exerted a cooling of about a quarter to half a watt per square meter (roughly a square yard). Without peatlands' consistent drawdown of carbon dioxide, the current warming of about 1 watt per square meter would be that much worse. The researchers made their estimates using detailed information on peat accumulation rates since the end of the last ice age, with Siberia, western Canada, and Finland representing northern peatlands as a whole.

Other analyses more specific to modern estimates of peat production versus methane releases suggest that peatlands may have a short-term warming effect in some years, possibly even for decades at a time. Like many other natural systems, though, northern peatlands have been helping to slow the pace of global warming over the centuries.

It's more challenging to consider how freshwater tropical peatlands balance out. They tend to be several times more productive than their northern brethren. But freshwater wetlands in the tropics also tend to release more methane, in part because their plants and soils don't face a killing frost that shuts down most emissions along with productivity. Some analyses suggest that tropical wetlands contribute half or more of the world's methane from wetlands, even though they account for a third or less of wetlands area. This suggests that wetland methane emissions relate more to total productivity than area of coverage. In fact, wetlands expert William Mitsch is reaching just such a conclusion, as he shared during a visit to his Ohio research lab.

Along these lines, Florida wetlands collected a third to twice as much carbon as systems in Canada and Virginia for the same amount of methane emitted, based on a comparative study by researchers Gary Whiting and Jeffrey Chanton. What's more, coastal wetlands release almost no methane. "Wherever you have saltwater, you don't find methane release. That's only in freshwater," explained Robert Twilley, a Louisiana State University professor who specializes in the biogeochemistry of wetlands. Mangroves, coastal marshes, and other systems intimately linked to the sea release sulfides rather than methane, he said. "That's why saltwater wetlands smell like rotten eggs. I tell my students they can always tell the presence of saltwater wetlands just by doing a sniff test." Researchers in Puerto Rico similarly concluded that methane emissions are "negligible" in coastal mangroves.

Overall, it seems likely that tropical and subtropical peatlands also cool the planet as a whole. For one thing, many of the peat-forming systems reside in mangroves or saltwater marshes, where methane production is minor or nonexistent. For another, methane degrades much more quickly in the warm tropical atmosphere than it does in the cooler air around the poles. In modern conditions, its atmospheric life span ranges from less than a decade in the tropics to about a quarter of a century near the poles. Its longer duration near the poles, however, theoretically could increase local warming above high-latitude peatlands, if only for the growing season before winds fully disperse methane concentrations.

Researchers such as Lisa Sloan continue to investigate how much methane emissions from wetlands might have contributed to keeping polar regions warm during hothouse climates. In the big picture, though, peatlands through geological time had a well-documented effect in reducing greenhouse-gas levels and thus cooling the planet overall.

❧

Wetlands have an intimate connection with plants and carbon dioxide levels through the ages, as Stephen Greb and colleagues elegantly described in the 2006 book *Wetlands through Time*, the source of much of the following summary. When mosses and lichens became the first plants to venture onto land some 450 million years ago, they started in wetlands. The first vascular plants—that is, those featuring stems—similarly emerged from wetlands, starting with *Cooksonian*, a small sticklike plant that favored floodplains around rivers. Even the oldest known tall, woody tree—*Archaeopteris*, an outgrowth of Devonian time—started its existence in riverside and coastal swamps. The deep, extensive root systems of these trees and those of their relatives formed the first swamps credited with reducing catastrophic flooding around wetlands. Devonian swamps also supported diverse wildlife. Fallen leaves nourished mud-dwelling creatures that launched a complex food chain. The first evidence of peat production comes from the coal beds of the late Devonian. Their effectiveness in storing carbon dioxide as plants were buried in waterlogged conditions led to a carbon drawdown linked to an extensive cooling. The drawdown began in the Devonian era and became precipitous with the spread of trees some 330 million years ago during the subsequent Carboniferous.

Vast equatorial wetlands thrived during the Carboniferous, which continued the general trend toward declining carbon dioxide levels and temperatures. This era earned its name from its vast coal deposits, most of which began their existence in tropical peatlands. The first great coalfields of North America, Asia, and Europe formed in tropical lowlands during the Carboniferous. These were generally wetlands bordering the closing Rheic Ocean as the two major landmasses moved toward each other. Wetlands that didn't form peat also thrived, as shown by the abundance of preserved plant parts. As a greater share of the carbon in their leaves and wood faced long-term burial, the airborne carbon dioxide levels declined. By about 290 million years ago, carbon dioxide levels had dropped to levels on par with recent ice ages. Predictably, global temperatures dropped, too. The climate settled into an 80-million-year-long ice age that spanned the Carboniferous and subsequent Permian. Here was an example of how Gaia's climate control system could metaphorically lurch from one extreme to the other even while keeping conditions within a livable range.

As the Carboniferous period drew to a close, swamp habitats dried out, leaving many water-loving plants high and dry. At the start of the icy Permian, wetlands had contracted on most continental plates. North American coalfields were restricted to the Appalachian basin, and tropical coals thinned out to only a few areas in Asia. The tropical belt of rainfall had narrowed over most of the interlocking continents. Where maritime conditions widened this belt, such as around several Asian plates including the one holding modern-day China, tropical and subtropical coals continued to build. But the widespread aridity over the conglomerate of continents known as Gondwana shut down tropical peat production there. Wetlands shifted toward the poles. The wetlands thriving in the cool, moist high latitudes included a first—peat accumulating under permafrost conditions. Elsewhere, however, wetlands generally suffered a decline during the Permian, even after the ice age ended. By then, Gondwana had moved up from its position around the South Pole to join up with all the other continents in a unified landmass known as Pangea, which stretched from the South Pole almost all the way to the North Pole. For its equatorial region, the joining of these landmasses meant large-scale drying in a continental effect. Wetlands contracted to riparian corridors and lakeside settings.

The loss of wetlands may have contributed to the end-of-Permian extinction event about 250 million years ago. Theories abound about why up to 90 percent of existing species disappeared during this interval, which apparently featured a sudden drop in oxygen levels as well as a rise in carbon dioxide levels, as Peter Ward describes in his 2006 book *Out of Thin Air*. At any rate, the event led to the almost total collapse of most wetland species, including many dominant trees that had favored peatlands during the preceding two periods spanning the Carboniferous, as Stephen Greb and colleagues suggested in *Wetlands through Time*. The vanishing wetlands remained out of the picture for many millions of years during the next time frame, the Triassic. The fossil record shows no evidence for coal development anywhere in the world for about 6 million years—about the same time coral reefs also were reduced to a fraction of their former range. In a 2006 exposition, Gregory Retallack and Evelyn Krull consider this one of the lines of evidence that suggest massive bursts of methane gas may have accompanied—perhaps caused—the warming that eventually pulled the planet out of the lengthy ice age. A greenhouse warming at the end of the Triassic occurred at the same time as a mass extinction event that included global changes in carbon dioxide levels and an abrupt turnover of most large plants.

During the Jurassic, further warming accompanied by an increase in precipitation allowed wetlands to expand again. By the end of the Jurassic, coal was accumulating in basins in China, Mongolia, Iran, and the former Soviet Union. By the subsequent Cretaceous, wetlands had grown from slender swamps and marshes fringing rivers and lakes to extensive swamps and mires. Another major era of coal production had begun, second only to the Carboniferous for sheer abundance. The early Cretaceous also marked the evolution of the angiosperms, a colorful clan containing every flowering plant and hardwood tree in existence. With the angiosperms came water lilies, lotus, and other aquatic floating plants that dropped their roots into the sediments below. Fens, bogs, and forest swamps widened their grip on the landscape. The flowering angiosperms thrived on much of the landscape, but coniferous trees continued to dominate swamps. Cedar, redwood, and dawn redwood (*Metasequoia*) dominated at higher latitudes. "Metasequoia was a very important component of wetland vegetation," explained Scott Wing, a researcher and curator with the Smithsonian's National Museum of Natural History. Dawn redwoods were ubiquitous in Europe, Greenland, eastern Asia, and other latitudes above 40 degrees, he noted.

The large quantities of Arctic coal from the Cretaceous represent an extremely effective carbon sequestration system. Winter dieback or deciduousness combined with high precipitation rates and a delta system that encouraged sediment compaction all led to the preservation of large amounts of organic matter. About a third of U.S. low-sulfur coal reserves come from Cretaceous deposits along the Arctic Slope. By the late Cretaceous, mangroves began to contain some of the species still found there today. Because of this, they probably had developed many of their modern functions: trapping sediments, reducing storm surge, creating oases for diverse species, and collecting and exporting nutrients.

❧

About 100 million years ago during the mid-Cretaceous, the landlocked Four Corners region, where Arizona, Utah, Colorado, and New Mexico square off, sported a seashore. An inland sea stretched from the Gulf of Mexico to the Arctic, covering parts of what is now the Colorado Plateau with water. Ferns intermingled with diverse softwoods such as monkey puzzle trees as well as some of the more recently evolved flowering plants and trees, including magnolias. These productive swamps and the nearby sea left behind a legacy of coal, oil, and gas, mainly on Navajo and Hopi land, that continues to store carbon today. Oil and gas form mostly in the sea, only when high pressure and high temperatures follow high productivity, a topic of the next chapter.

Coal forms under much less stringent conditions. Basically, anywhere peat grows, coal could eventually form. Its formation requires higher pressure and temperature than found on the surface, but millennia of peat production can be enough to compact the underlying material into coal veins. As new plant material buries the old organic matter, water gets pressed out. It starts to compress into lignite. Note the tip of the hat to wood lignin in that word. That's because woody material, whether in roots or logs, is the most likely to resist decay long enough to become peat. Further compaction brings harder coal, culminating in anthracite, the diamonds of the coal world. Whether the end result is lignite, anthracite, or an intermediate variety, coal formation invariably goes back to wetlands.

❧

Through the ages, wetlands have tended to increase in areas or times when precipitation rates are relatively high. Judith Totman Parrish and her colleagues looked at half a dozen stretches of deep time to consider what makes coal-producing wetlands thrive. They found that coal bearers of the past tended to group around either the equator or that sweet spot favored by modern wetlands, around 55 degrees North and South. This region of the globe often receives moisture from Hadley circulation during relatively warm climate regimes, including our modern one. The time slices the researchers considered across the 250-million-year time frame included several phases when the temperatures at the 55-degree latitude were quite a bit warmer than now. Like the Cretaceous, the Eocene hothouse also produced extensive coal beds, such as in the Eocene Green River Formation, when inland seas helped keep parts of Wyoming and Utah warm and wet.

In the more recent past, more of the world's peat production occurred during interglacial warm periods rather than ice ages. For instance, carbon-dating of the base of northern peatlands consistently puts them at less than 10,000 years old. In other words, they arose only after the last glacial age drew to a close. That's not surprising given that extensive ice sheets covered most of the area featuring these productive northern peatlands for more than 100,000 years before the latest meltdown. It's a safe bet that much of the last 2 million years featured similar cycles, with northern peat production thriving during interglacials, then grinding to a halt during glacial climates as ice sheets plowed over them.

Through the ages, the creation of peat and coal has worked as a long-term means of storing carbon during warm, wet periods. It worked well, anyway, until our ancestors figured out how to extract and burn the stored material. The peat, coal, and oil produced during earlier warm epochs provide a major source of the fossil fuels we use in modern times. "Coal is really the problem. We can get ourselves into trouble with oil and gas, but we can kill ourselves with coal," Meinrat Andreae, a director at Germany's Max Planck Institute for Chemistry and an expert on air pollution, explained during a 2008 talk hosted by the University of Arizona's Department of Soil, Water and Environmental Science. Burning all the world's oil and gas could double the amount of carbon dioxide in the air, whereas burning all the coal would quadruple it, he said. The much easier geological transition to coal versus petroleum products helps explain why we have more coal on this planet than oil and gas.

Wetlands and their plants are at the root of coal production through the ages. Meanwhile, offshore plant productivity can create the conditions that lead to the formation of oil and gas. In warm times of high seas, the amount of land covered by wetlands and shallow seas increases (as chapter 4 described). While the expansion of wetlands promotes the formation of peat and coal, the abundance of shallow seas promotes the coastal productivity that can create oil and gas products. The process leading to oil and gas formation starts where plant productivity overwhelms the marine system's ability to decay it—which also occurs more frequently during warm, wet climates. And the production of oil and gas, along with the formation of limestone and other carbonates and the weathering of rock, draws down carbon dioxide over long time frames.

7

Beneath the Surface
Weathering the Warming in Deep Time

During a cloudburst in Puerto Rico's Luquillo Mountains, a three-foot-wide stream sprang up between my partner, Bob Segal, and me. At the time, we were measuring tree diameters during a 1998 survey of Hurricane Georges damage. Although he was only about four feet away, the newly formed stream was moving fast. We each grabbed the nearest tree to keep ourselves from sliding down the steep clay-covered slope while we waited out the quarter-hour downpour. The ephemeral stream between us carried visible leaf material along for the ride, as well as the brownish-orange clay lifted from local soil stores. On another memorable occasion a few weeks later, a heavy rain trapped us on the steep slope of a hill denuded by a landslide, which turned up in one of our plots in Cayey. We were directly upslope from a rushing river, now brown and frothy from the sediment and water washing into it. Several uprooted trees stood as sentinels between the slope and the river, their roots already cleaned of soil by other storms since their fall two months earlier. After about half an hour, when the rains failed to let up, we inched along a trail of fallen trees onto safer ground, taking care not to join the soil particles and debris being pulled off the mountain into the river.

Anyone who has observed swiftly flowing rivers, whether in person or on the news, will recall that the rushing waters churn with shades of local soil—whether muddy brown or dirt red, whether in the Yellow River or the Río Negro. Hurricane Katrina floods dropped four to eight inches of mud as the waters settled in Louisiana. A July 2006 flood in Norwalk, Ohio, similarly left the aptly named Water Street covered with at least an inch of mud after the overflowing water had drifted back into the nearby creek. "It smelled like the river bottom—stinky," said Edward Christie, a service technician with a local business. Other U.S. floods that year brought mud to homes around the Sandusky River, the Delaware, and various eastern rivers. A late July 2006 flood in Tucson's

Sabino Canyon carried so much material that it clogged culverts under roads, then destroyed overlying bridges when it jumped its banks. Mississippi floods regularly make the news, with 2008 overflows being among the most recent and the large-scale 1993 event being among the most memorable.

These floods offer a fast-forward glimpse at an ongoing process that has a slow but definite impact on global carbon dioxide levels: weathering. Intense rainfall speeds up the rate of both physical and chemical weathering. For simplicity, I'll refer to the former as erosion and the latter as weathering. Although both move mountains, weathering and erosion differ slightly. Weathering turns bedrock into soil. Erosion moves the soil off the slopes. Dusty winds can erode soft material such as sandstone, and ice can help split soil and rock. Other than that, both erosion and weathering rely on the presence of liquid water to do their jobs. Weathering plucks carbon dioxide out of the air as it works at the molecular level to divide rock into smaller parts, including chemical compounds such as bicarbonate and calcium. Then it carries some of those compounds off into rivers and eventually the sea. Erosion whisks off soil, along with the organic carbon it contains, as well as leaves and other carbon-rich debris. Rivers then deposit some of that carbon into wetter environments where it is less likely to decay.

The organic matter traveling by river that makes it to the seafloor without being eaten or dissolved can put former carbon dioxide into deep storage. Ocean sediments hold several times the amount of carbon in the atmosphere. The ocean itself holds about fifty times more carbon than the air, counting dissolved bicarbonates and carbonates. Coal and oil deposits of the world, meanwhile, hold at least several times more carbon than the amount in the air, while the mostly inaccessible but somewhat unpredictable natural gas of the sea, methane hydrates, contains more than ten times the amount of carbon residing in the air. The biggest storehouse of marine carbon, though, resides in solid carbonates—limestone, dolomite, and chalk, mainly—many of which owe their existence to coral reefs and to microscopic shell-forming creatures known fondly as forams and coccoliths. By land and by sea, carbonates hold about quadruple the amount of carbon found even in the world's methane hydrates deposits. That's about 42,000 billion tons of carbon, according to Peter Skelton and the other authors of *The Cretaceous World*, roughly sixty times the amount in the air. This points to the importance of the weathering process, at least over the long term.

Gaia appears to employ weathering to balance out excursions of carbon dioxide over millions of years. This is the method of carbon dioxide drawdown most often considered by scholars of both Gaia theory and Earth system science. Although these processes take longer than we can wait if we want to avoid city-obliterating sea-level rises and bigger hurricanes and floods, they are worth touching upon here to consider how Earth balances its carbon dioxide ledger in deep time. Along with the weathering process itself, this chapter will cover some marine deposits created with assistance from weathering: coral reefs, and oil and gas.

❦

The sediment, logs, and leaves carried to the coast during floods eventually would leave the landscape barren were it not for the ongoing weathering that converts the underlying rock into soil. Weathering occurs when water seeps into bedrock, often toting carbonic acid—the liquid form of the carbon dioxide it collected on its journey through the atmosphere and soil. Nature's version of acid rain helps break the bonds making rock solid. Raindrops carrying carbonic acid dissolve limestone deposits, micrometer by micrometer, carrying molecules of calcium and bicarbonate down to the sea. In the case of limestone and other calcium carbonate formations, the chemistry involved means the carbon dioxide taken down via weathering eventually escapes back into the atmosphere. It may take hundreds of thousands of years to make its getaway, but on million-year time scales no storage results. On the other hand, the weathering of basalts, granites, and other silica-based rock pulls down some carbon dioxide into a life sentence, in geological terms. It's the molecular equivalent of cement shoes—a burial in this substrate will generally hold down the formerly free-floating carbon dioxide until the deposited carbon becomes fodder for an active volcano. So the weathering of silicate rocks gradually reduces greenhouse-gas levels over the ages. As it happens, weathering speeds up in hot, humid climates, thus pulling more carbon dioxide out of the air than during cool, dry times.

"Weathering comes from weather, for Pete's sake. You can't have weathering in the middle of the desert—there's no rain," explained Robert Berner, a longtime geology professor who merited a special issue of the journal *Geochimica et Cosmochimica Acta* when he retired from Yale University in 2006. Many of his scientific papers have addressed the various influences and results of weathering over the past 500 million years, as does his 2004 book *The Phanerozoic Carbon Cycle*. In between writing scientific publications, Berner plays and composes music for piano and other instruments. Like the ups and downs of his elegant musical numbers, the models he has developed capture scale changes—in this case, the long-term fluctuations in levels of carbon dioxide and other atmospheric gases. Even as his models grew more sophisticated over the decades, Berner found that estimating levels of this greenhouse gas over the eons largely boiled down to modeling weathering rates. The results generally reflect reality, as witnessed by their comparison to physical evidence compiled in an analytical review by colleague Dana Royer. (Admittedly, the estimates leave plenty of wiggle room for the distant past, given the long time spans considered and the challenges of assessing gas levels in vanished thin air.) A few fall outside the model range, but most of the hundreds of data points drawn from marine sediments, ancient soil nodules, and other techniques fall squarely within Berner's estimates.

In the 1980s, Berner and a couple of colleagues proposed that the factors that come with a greenhouse climate—including higher temperatures, more rainfall, and the trees that flourish in these conditions—work together to pull down carbon dioxide. This perspective became known as the BLAG hypothesis, for the last names of its authors (Berner, Antonio Lasaga, and Robert Garrels). At the 100-million-year time frame they cover, volcanoes control carbon dioxide emissions. These fiery mounts go into a relative frenzy when the seafloor spreads more rapidly than usual, as they did during the Cretaceous. (Changes in Earth's magnetic field allow dating of the progression of hardened magma around sites where the seafloor is spreading, yielding estimates of volcanic activity during each epoch.) Carbon dioxide represents about a fifth of volcanic emissions, on average, with water vapor making up most of the remainder. Weathering rates, in turn, rise in response to the rising levels of moisture and carbon dioxide levels, as well as the greenhouse conditions they create on the planet.

Like the sediments that are visible to observers of floods, differences in weathering rates also show up at much shorter time frames. In modern times, bedrock in the humid tropics weathers roughly eight times faster than comparable bedrock around the cold, relatively dry poles. Researchers made these estimates using the chemical products of this process carried by rivers. Temperature and the amount of water flowing on the landscape drive weathering rates of basalt, a silicate, based on research led by Céline Dessert of France and earlier work considered in her analysis. Granites weather more slowly, but their weathering rate, too, speeds up in warm, moist climates. Evidence in the geological record supports the case that weathering rates increased during warmer climates featuring elevated carbon dioxide levels, as geologist Gregory Retallack points out in a 2001 *Nature* paper. For instance, he noted that soil development during the hothouse Eocene occurred more rapidly than during the cooler Oligocene. Soil develops as bedrock weathers into grains.

Berner's "favorite lecture" and the topic of several of his scientific papers involves the role of plants on weathering rates. "It's mainly trees we're talking about, not just any old plant. Trees do the best job of weathering because they have extensive root systems," Berner said. He and a colleague, Katherine Moulton, found that modern-day weathering rates under trees growing in Iceland are three or four times faster than the rates for neighboring ground without trees facing similar environmental conditions and published the results in 1998 with Moulton as lead author. Other research has estimated that plants increase the rate of weathering by two to ten times, with the higher rates counting the storage of weathering products such as calcium in the soil and in the trees themselves. (Calcium, the same material that hardens bones, gives strength to tree trunks.) Using this information and other factors, Berner estimated that carbon dioxide levels about 450 million years ago were about fifteen times higher than recent preindustrial levels. Plants with penetrating roots evolved about 380 million years ago, during the Devonian, about the same time swamps first appeared. Earlier plants, such as lichens, have a limited ability to weather rock. "I've seen lichens on ancient Roman graves where you can still read the names," Berner said. Plants, especially trees, release organic acids that assist the weathering of bedrock into soil particles. Their roots can also widen cracks in rocks.

The accelerating effect plants have on weathering rates is the method of greenhouse-gas regulation most acknowledged by Gaian thinkers. For instance, in a 1989 *Nature* article, David Schwartzmann and Tyler Volk crunched some numbers to support the hypothesis that the planet would have far less soil clinging to its surface without the intervention of plants. First, plant roots jack up the levels of carbon dioxide in the soil by ten to one hundred times atmospheric levels, thus boosting the amount of carbonic acid that can chip away at bedrock. This alone probably at least doubles the weathering rate and may increase it sixfold, they suggest. Plants also release their own acidic compounds including organic, humic, and fulvic acids. The plants with the highest productivity—generally trees—produce the most acids. Forests release an order of magnitude more acids into the soil per square foot than ecosystems limited to algae and lichens, another analysis concluded.

Plants also help stabilize the resulting soil particles, Schwartzmann and Volk point out. They bind particles with their roots and the organic matter they exude. Dead roots, dropped leaves, and other decaying plant parts add organic matter to soils. Soil organic matter helps soil particles stick together. Without it, soil grains can act like shifting sands. By keeping the soil from drifting off during winds and rains, plants also maintain a three-dimensional matrix of particles for exposure to the ravages of weathering. Bare rock, in contrast, exposes only a flat, two-dimensional surface to weathering. Finally, the soils plants create and maintain increase the landscape's ability to store water, thus increasing the amount of time bedrock will come face-to-face with water and the carbonic acid it holds. By stabilizing soil and increasing the area of potential contact with the forces of weather, plants probably enhance weathering rates by a hundred to a thousand times, Schwartzmann and Volk concluded. Without soils, "the lowlands could be virtual deserts with most rainfall concentrated in mountain ranges," they state.

The ability of trees to boost weathering and otherwise draw down carbon dioxide has led some scientists to suggest that their evolution may have launched a long-term decline in carbon dioxide levels associated with the ancient Carboniferous-Permian ice age (summarized in the previous chapter). The conversion of carbon dioxide into the carbohydrates comprising trees made a dent in carbon dioxide supplies. The lignin formed by trees remains resistant to decay even today. It's likely that even fewer creatures knew what to do with this carbon-rich material when it first evolved. The burial of lignin in swamps and on the seafloor over time put billions of tons of former carbon dioxide out of commission. Similarly, the rise in hardwood trees during the Cretaceous has been linked to a decline in carbon dioxide levels from the mid-Cretaceous through the early Eocene, until a burst of methane gas apparently bumped levels of greenhouse gases back up.

The drawdown in carbon dioxide since the Eocene, meanwhile, has been linked to the higher weathering rate that comes with mountains. Work by Maureen Raymo alone and with colleagues suggests that the overall global cooling since early in the Eocene traces back to the creation of the Himalayan Mountains and the Tibetan Plateau about 40 or 50 million years ago. As the formerly freewheeling continent of India slammed into Asia, the land crumpled like a piece of tinfoil into the peaks and valleys of the world's highest plateau. When these compelling mountains rose from contorted land, their steep slopes increased the amount of sediment and weathering products moving into the Indian Ocean. Although it remains unproven, many scientists support the theory that the weathering of these mountains helped launch the return of polar ice caps. Today, the heights of the Tibetan Plateau help keep Asian rivers carrying about 70 percent of the world's sediment load to the sea, where some of its carbon faces long-term burial in the sediments of the Bengal Fan. Incidentally, recent research suggests that the Andes Mountains rose abruptly about 8 million years ago, in the Miocene—right about the time low-carbon (C_4) plants evolved, arguably in adaptation to declining atmospheric carbon dioxide levels. (Chapter 5 discusses low-carbon plants in more detail.) Over time, the products of weathering washing off mountains and hillsides nourish coral reefs, support the marine food chain, and stabilize sea chemistry. Mountains everywhere tend to bump up weathering rates, especially where they are boosting local rainfall and supporting forests as well.

The presence of mountains speeds up the weathering rate in several ways. For one thing, mountains promote precipitation (described in chapter 4). Although temperatures tend to cool with height up a mountain, the extra precipitation mountains foster overrides the declining temperature in many cases or seasons. This promotes chemical weathering. Mountains promote erosion and weathering, too, because of their angles. Steep slopes bring on higher erosion rates, with soil peeled off millimeter by millimeter. This influence becomes apparent to hapless bystanders and researchers during rainstorms on steep slopes. So, for instance, studies have found that the Andes have much higher weathering rates than the nearby tropical rain forests. The soils under the rain forest already have been weathered thoroughly, with their particles turned into fine clay and many of their nutrients washed away. To continue weathering, percolating water often must trickle through dozens of feet of fine clay before reaching bedrock. Meanwhile, steep slopes can drop millions of cubic feet of soil in one slump during major landslides. This clearing of topsoil keeps bedrock closer to the surface, where it is more easily reached by the ravages of weather. The additional weathering, in turn, provides more fodder for coastal reefs.

Beneath the sea surface of many coastal regions, rays of sunshine dapple coralscapes of color. Reefs have been called the rain forests of the sea because of their high levels of productivity and biological diversity. Yet in some ways, they bear more resemblance to wildflower season. Reds, purples, and yellows stand out among the grays and whites, in a stunning variety of shapes. From cauliflower-shaped clumps to lacy webs, they all fit into the coral clan. The seeming delicacy of some of the designs belies their resilience in the face of storms. Color blooms on the reefs year-round, when they're healthy. When they're overheated and otherwise stressed, they can take on the pale coloring of a plague victim.

Coral reefs in modern times have been facing a series of assaults, including "bleaching" episodes that occur during times of higher-than-usual water temperatures. It takes only a month of sea-surface temperatures roughly 2 degrees Fahrenheit above normal to cause symbiotic corals to "bleach," explained Richard Feely, supervisory oceanographer at the National Oceanic and Atmospheric Administration's Pacific Marine Environmental Laboratory in Seattle, during a talk at the University of Arizona. They do this by expelling most of their algae, he said, leaving the remaining bleached-out coral without its life-support system. Sometimes algae return to dwell in the coral once again. Or a new, more heat-tolerant species may move in. All too often, though, the algae never return. Then the formerly thriving coral reef can turn into a ghost town. With no algae around to make repairs and encourage growth, the corals can start to decay as waves erode their surface.

Another potential plague on the world's reefs involves ocean acidification, which Feely referred to as "global warming's evil twin." The concerns over acidification relate back to carbon chemistry in the oceans. In water, carbon may take several other forms besides carbon dioxide—with or without the intervention of life. Upon reacting with water, seaborne carbon dioxide can transform into carbonic acid. This is the same mild acid that helps boost the weathering rate of rocks on land. Or it can morph into bicarbonate or even carbonate. These chemical changes involve the release of positively charged hydrogen ions into the surrounding water in the process. The hydrogen ions involved in these transformations help define the acidity of the water, thus affecting its pH. The pH, in turn, helps determine which of these inorganic carbon forms dominates the chemical stew of compounds. Ocean pH tends to run slightly alkaline. But researchers have measured a drop in ocean pH from about 8.2 to 8.1 since preindustrial times, Feely said. This shift comes with a slight drop in the saturation levels of the calcium carbonate compound aragonite, the main component of corals.

Weathering helps balance ocean acidity. Much of the free-floating sediment so visible during floods eventually reaches the coast. Rivers also carry dissolved compounds so small they don't block the sun reaching the reef. These weathered compounds include magnesium and calcium leached from the rock matrix, whether limestone or silicates. Much like the Milk of Magnesia or calcium-based Tums we might swallow to combat an acid stomach, the magnesium and calcium flowing to the sea help neutralize the carbonic acid arriving from various sources, including the air. Even more important in balancing ocean acidity is bicarbonate. The powdery form of sodium bicarbonate is a main ingredient in traditional baking soda. In its dissolved state, it buffers acidity. It often morphs into carbonate, a main building block in the construction of reefs and shells. Calcium, the same element that builds bones and tree trunks, is the other main compound that gives structure to reef communities.

Rivers carry bicarbonate, calcium, magnesium, and other bases captured in the weathering process into the ocean to buffer it from acidity in times of high carbon dioxide levels. But there's a catch: The balancing act appears to work at the scale of thousands of years. Berner kindly did some calculations in response to my question of how much one year's worth of riverborne weathering products might offset the ocean's absorption of airborne carbon dioxide released by humans. The answer was a fraction of 1 percent, a number that suggested it would take about 500 years of weathering contributions to balance out one year of carbon dioxide absorption. The ocean has some extra buffering in its system now, but not enough to offset another century of escalating fossil-fuel emissions. That's why many scientists worry that the ocean's acidity could be thrown off-balance over coming decades. It's happened in the past, such as when an undersea methane release helped launched the Eocene (described in chapter 3).

The shape-shifting skills of carbon in the ocean make it difficult to predict just what this essential molecule will do in the future ocean. We are simultaneously changing the chemistry of not only the ocean itself, but also the air it traps in curling waves and the land that feeds it a watery soup of compounds. The greenhouse gases we've unleashed are changing the ocean temperature as well as its chemistry. These gases have their own measurable and yet somewhat unpredictable effect on ocean dynamics, chemistry, and the currents that affect both. It seems likely that the colder sections of ocean, such as the Arctic Sea, will be the most dramatically affected by the acidity issue, Feely indicated. A look back at how reefs handled the hothouse, high carbon dioxide environment of the Cretaceous will help illustrate some of the issues.

❧

For many years, the occurrence of calcium carbonates in their various marine forms—coral reefs, oozes, chalks, and minute life-forms such as coccoliths and forams—was seen as an oceanic method for sequestering carbon dioxide. After all, the carbonate locked up in reefs and other shelly material is out of the atmosphere's reach. But, again, the issue is more complex than a first glance suggests. That's because the process of joining calcium and carbonate together into shelly compounds typically *releases* carbon dioxide. Weathering of land-based calcium carbonates, such as limestone cliffs and chalk deposits, occurs roughly ten times faster than weathering of silicates, as mentioned above. Yet only weathering of silicates takes down airborne carbon dioxide for the long haul. That means only about 10 percent of the weathered carbonates entering the ocean is likely to contribute to the ledger of carbon dioxide uptake over the longest time frames. Even so, hundreds of thousands of years might pass before the cast of characters from either carbonate or silicate weathering transforms into carbonates. Perhaps if the broken shells weathered off of limestone cliffs and chalk deposits weren't being put back together again as calcium carbonate in the sea, they could have stayed in cold storage even longer. But the construction of reefs and shelly microbes launches half of them back into a gaseous form, giving them the opportunity to float off and resume work as a greenhouse-gas molecule. The other half converts into calcium carbonate.

The joining together of calcium and carbonate is more likely to happen in warm times with warm, shallow seas. It's not even always necessary to invoke life to explain why warmer seas are better at converting dissolved compounds into solids. For instance, high temperatures around the Bahamas can encourage the nearby sea to face summertime "whitings," when formerly dissolved calcium and carbonate molecules join together in visible solidarity. Why? Whitings occur when there's an oversaturation of these dissolved compounds for a given water temperature. Other things being equal, carbonates are more likely to precipitate—in this case, change from an aqueous to a solid state—as water temperature rises. It's almost the reverse of how precipitation of water vapor works. Warm air holds more moisture, or water vapor (as chapter 4 described). So if that air cools off, some of the water vapor inside will condense back into liquid water. That's the basic impetus behind rain and snow. Meanwhile, cold water holds more dissolved carbonates. Below a certain temperature, which varies depending on salinity and saturation levels, carbonate cannot exist in solid form. In fact, all oceans deep enough to feature cold subsurface waters contain a shifting boundary known as the carbonate compensation depth. Above this imaginary boundary, carbonates can exist in solid form. Below it, they dissolve back into aqueous compounds. This shifting boundary may be twice as deep in the tropics as around the poles.

Local warming of shallow continental shelves may help explain why warm climates of the past often featured widespread carbonate platforms in tropical coastal waters. The expansion of coastal reefs during past interglacial warm periods is well known. The sea-level rise permitted reefs to expand well beyond their ice-age locations. In fact, back in the 1980s, Wolfgang Berger invoked these recurring reef expansions to try to explain why carbon dioxide levels reached the heights they did in the ice-core bubbles capturing interglacial air. Later work, however, pointed out that the timing wasn't quite right for coral reefs to trigger the rise in carbon dioxide levels. Detailed analyses show that the carbon dioxide levels generally rose *before* the ice sheets melted (chapter 3). Ditto the sea-level rise, by extension. So carbon dioxide must have reached its interglacial heights before the seas cultivated additional coral reefs. Still, it seems possible that the reef construction helped maintain relatively high carbon dioxide levels despite the faster weathering rate that comes with the warmer climate. A similar argument has been made regarding the Cretaceous hothouse, a period of abundant carbonate production with carbon dioxide levels skyrocketing three to eight times higher than the usual interglacial level. On the other hand, there's no denying that some carbon gets locked back up into a solid matrix when reefs form, as they are more likely to do in warm climates.

"Modern shallow marine carbonates are now restricted to a few pathetic ghostlike remnants of a once extensive distribution," geologists Donald Prothero and Fred Schwab stated in their 1996 textbook *Sedimentary Geology*. Like most geologists, they lump our modern climate in with other relatively cold phases that featured extensive polar ice sheets. The icier the age, the colder and smaller the seas. As seawater recedes from continental shelf areas, shorelines can sink to the depths of drop-offs. In waters deeper than several hundred feet, the sunlight filtering down to the seafloor is too sparse to support the symbiotic algae that build coral reefs. Colder waters also lean toward dissolving carbonates, making it more difficult for them to retain their solid forms. Meanwhile, the shift to a cold, dry climate brings a decline in weathering rates, meaning less calcium and less bicarbonate arrive in coastal waters.

The level of reef expansion during the hothouse Cretaceous even went beyond the expansion during past interglacial warm periods. The high temperatures, high carbon dioxide levels, and vast rivers of the Cretaceous increased the weathering rate. High sea temperatures, meanwhile, boosted the formation of extensive calcium carbonate platforms in shallow coastal waters. In fact, the widespread presence of calcareous chalks during the Cretaceous helped the period earn its name: "Creta" means "chalk" in Latin. By about the mid-Cretaceous, coccoliths and forams had evolved in the world's oceans, signaling their arrival with chalky oozes formed as their calcium carbonate shells littered the seafloor. The abundance of these chalky deposits also indicates the widespread extent of coastal waters, as calcareous ooze accumulates only in relatively shallow depths. A plentiful group of coccoliths, golden brown algae, created extensive carbonate platforms in waters up to 600 feet deep. This expansion into waters deeper than what coral reefs could handle opened up what Peter Skelton and his co-authors, in the book *The Cretaceous World,* call "an important new pathway" for pulling down carbon dioxide on long time scales.

Along with chalks, tropical regions also featured Cretaceous limestones, mainly derived from an extinct group of clamlike creatures known as rudists. They episodically developed carbonate platforms. In truth, these so-called reefs were quite different from modern reefs. To an observer, they might have resembled an oyster bed more than the epicenters of biodiversity and productivity we encounter today. A platy version of coral reefs lived in slightly deeper waters offshore. In the warmest, saltiest waters of the mid-Cretaceous, though, rudists often replaced the more diverse symbiotic corals. Reef-building corals also ventured closer to the poles during this hothouse period. Coral reefs thrived at 55 degrees North, about the latitude of the Alaskan peninsula, whereas modern reefs won't really venture past 37 degrees North (roughly on par with San Francisco) around the globe, as Judith Totman Parrish notes in her 1998 book.

Ancient reefs also faced times of episodic destruction during hothouse periods, however. There's evidence of widespread drowning of reef and carbonate platforms during the mid-Cretaceous and other hothouse periods. The "drowning" description is a bit of a misnomer. Reefs usually can keep up with sea-level rise, geologists figure, so it's more likely that they were overly inundated with nutrients or sediments rather than with water. Given the voracious appetite for rock that weathering processes demonstrated in the Cretaceous, it seems reasonable to suspect that too much sediment and their hitchhiking nutrients would arrive from the land from time to time. Even though apparently few mountains existed to give the rates a real lift, high temperatures, recurring storms, and abundant rivers exposed the land to plenty of weathering. Warm waters hold less oxygen, too.

Whether "drowned," buried, or starved of oxygen, many carbonate platforms clearly perished en masse across the seas on several occasions, as Peter Skelton and colleagues describe in *The Cretaceous World*. Right at the mid-Cretaceous about 100 million years ago, the carbonate platforms around the Americas and Caribbean started a sharp decline from a Cretaceous high that had peaked roughly 5 million years earlier. The reefs around Africa, Australia, and Europe hit their peak a bit after the decline of the "New World" carbonate platforms but joined them for an even more precipitous drop in a sudden event roughly 94 million years ago. Given the million of years considered at a stretch, it's difficult to see the details. An earlier episode has since resolved into several smaller events, for instance. Clearly, reefs were going through bust-and-boom cycles throughout the Cretaceous, but it's unclear why. Some of the reefs that died, though, received a layer of algae as their funeral shroud. While algae and other plants seemed to have a feast, their burst of productivity apparently stagnated the ocean enough to kill some carbonate reefs—sometimes many at once.

The fact that reefs thrived during previous interglacial warm periods and the Cretaceous suggests that warm temperatures alone do not lead to widespread dissolution of carbonate systems. Other factors besides temperature likely contribute to the modern problems plaguing the vortexes of ocean productivity that we call coral reefs. To explore more fully the role of carbonates in warm climates would take another book rather than a chapter section. In addition, their role in taking down carbon dioxide is limited by their tendency to release carbon dioxide during the construction process.

One study estimated that modern coral reefs emit the equivalent of about 1 percent of the carbon dioxide released in a year's burning of fossil fuels. Some researchers have measured carbon dioxide accumulating in the air over modern reef systems. Yet others have concluded that certain reefs were taking up carbon dioxide. Researcher Jean-Pierre Gattuso and a couple of colleagues tried to reconcile the latter observation with the understanding that the process of forming carbonate should release carbon dioxide. In a 1999 paper, they noted that the studies showing uptake of carbon dioxide tend to involve reefs fairly close to shore, with its many human impacts that can promote the growth of photosynthesizing algae over carbonate-building corals. The issue carries over to free-floating microscopic creatures as well. For instance, blooms from microscopic life-forms that build shells of calcium carbonate, such as coccoliths, consume less carbon dioxide than blooms of free-floating algae. (On the other hand, their extra weight may help them sink faster, thus making these shell-bearing plankton potentially more likely to carry carbon down with them in a deep-sea burial, Gattuso pointed out in a follow-up e-mail.)

Productivity among the phytoplankton that don't make calcium carbonate shells has a more clear-cut role in drawing down carbon dioxide for storage in both the short and long terms. The productivity of plants at land and sea during earlier warm periods clearly led to the long-term storage of former carbon dioxide as oil and gas. On the scale of human generations, though, the deadly algal blooms that create carbon drawdowns take many sought-after delicacies with them. Modern "dead zones" illustrate what happens when algae inundate large areas of coastal waters.

❧

In the summer of 2002, seafarers harvesting crabs off the Oregon coast pulled up pot after pot full of dead or dying crabs. They alerted Oregon State University researchers. At first, researchers figured it was an unusual event. But it came back, year after year, always during summer. Scientists and fishery managers who recently used an underwater video camera to scan the seafloor off Newport, Oregon, found a lifeless area covered with a sheet of ghostly white. The white mat turned out to be bacteria, part of an algal bloom that had depleted oxygen levels in an area that reaches the size of Rhode Island in some years.

Heavy winds can promote a short-term upwelling of nutrients that encourages the growth of marine algae, followed by a death-and-decay cycle that depletes local oxygen levels enough to kill slow-moving marine life. In some situations, such pockets of high-nutrient, oxygen-depleted conditions can assist a Gaian drawdown of airborne carbon dioxide.

Temporary algal blooms after hurricanes may represent one of these situations (as described in chapter 1), but these post-hurricane blooms are more of a flash in the pan. In coastal Oregon's case, the upwelling episode extends for months, typically from June through October. Through at least 2009, the dead zone was an annual feature. When Francis Chan and his colleagues—including Jane Lubchenco, who was appointed in March 2009 to head NOAA—measured the dead zone in 2006, it stretched across about 1,200 square miles of the continental shelf, from about 75 miles south of the Columbia River's mouth (at the border of Washington and Oregon) to nearly the latitude of Eugene. Chan and his colleagues theorized in a 2008 *Science* paper that unusually strong winds, possibly linked to global warming, had increased the amount of upwelling of deeper waters into the shelf region. Deeper waters tend to hold more nutrients but less oxygen than surface waters.

Although research into the causes of the Oregon dead zone has only just begun, scientists have convincingly established the cause behind an even larger dead zone that lurks off the coast of Louisiana—right about where the Mississippi River empties into the Gulf of Mexico. It, too, blooms during summer, expanding its deadly reach across some 8,000 square miles. It turns out that the time and place of the dead zone are no coincidence.

"The dead zone off Louisiana is fueled by the nitrate that comes down the Mississippi River that originates from agriculture, especially corn production, in the interior," as Robert Twilley, a Louisiana State University professor who directs the Coastal Sustainability Agenda, told me. Fertilizers draining from spring plantings taint the river's freshwater. Warmed by concrete channels on the way, the layer of freshwater comes to rest on top of the saltier—and thus heavier—seawater of the gulf. The algae that feast on incoming nitrates and other nutrients soon die and decay, depleting the oxygen in the stratified top layer when they do. By the time the last of their remains sink to the bottom, the lack of oxygen has suffocated slow-moving creatures that might otherwise have nibbled on them. As in the Pacific, the Gulf dead zone manifests in summer, when higher temperatures push out seaborne gases, such as oxygen and carbon dioxide, like the fizz in a warm soda. So, warm seas reduce oxygen availability even as fertilizer-fueled algal blooms are reaching peaks of productivity.

When such pockets of productivity become a recurring feature in the seascape, carbon comes down on a regular basis. In the really long term, these pockets sometimes lead to the creation of oil—a major drawdown of carbon dioxide at geological time scales. Although debate continues over which factor was more crucial in forming oil—excessive productivity or oxygen-depleted conditions—observations of modern dead zones suggest that these two factors usually go hand in hand.

❧

While coral reefs form out of inorganic carbon weathered from rocks, oil and gas deposits take shape from organic carbon—carbon products formed by life. Rivers carry at least as much organic carbon as inorganic carbon, perhaps twice as much. The riverborne organic carbon reaching the world's oceans amounts to perhaps 10 percent of the fossil-fuel emissions of a typical year. Just how much of this ends up stabilized in the sea remains a topic of debate. But clearly some of this organic carbon represents a long-term drawdown of carbon dioxide as it settles on the ocean floor. Meanwhile, the rivers' gifts of nitrogen and phosphorus enliven the marine community, nourishing the algae that appear to be the original energy source of oil and gas.

Where rivers pour their rich stew of carbon, calcium, nitrogen, and other life-supporting ingredients into the sea, productivity soars. Ocean productivity is highest "right where the water meets the rock meets the air," as Lynn Margulis put it during a talk at a 2006 conference on Gaia theory in Arlington, Virginia. Despite the much smaller area covered by coasts compared to the open sea, much of the ocean's "net primary productivity" (a measure of the movement of carbon from the atmosphere into green plants) is concentrated along the coasts. Algae have a lot to do with this. They form symbiotic partnerships to help build corals. They float around as plankton, building their tissues from carbon derivatives and other nutrients in the sea. They serve as the main course for other life-forms, big and small. Some algae can even get impressively large. Kelp, the seaweed that flourishes in shallow waters below about 80 feet, holds the size record for biggest brown alga. And in times and places of especially high productivity, algal remains can sink to the bottom of the sea.

The relatively rare combination of conditions for oil and methane production must also include a sealed burial, if humans are to find any petroleum puddles later. So only about 1 percent of organic carbon deposited in the sea ends up transforming into oil. Before becoming the hydrocarbons so beloved by our fuel-propelled culture, the long-dead algae and other ancient organic matter go through a transition stage as kerogen. This amorphous, organic-rich substance has faced enough pressure and heat in its long lifetime to make its origins rather ambiguous. But based on the content of oil shale—soft sedimentary rock that hasn't quite made the grade to become oil—algal blooms can play a big role in their formation.

The Eocene Green River Formation contains rich kerogen deposits in alternating layers that suggest seasonal bursts of algal productivity. These blooms apparently extended for hundreds of miles across the shallow waters that covered parts of Wyoming, Colorado, and Utah. The organic deposits that serve as the predecessors for oil might look like algal blooms in the ocean, too, agreed Judith Totman Parrish in response to my question. "But you need to have a system where you have sustained productivity," added Parrish, a biologist and geoscientist whose work on climates in deep time has encompassed fossil-fuel formation. The waste products of minuscule animals feasting on the algal blooms added organic matter to the mix. "There's such an abundance of food that zooplankton are just taking up whatever they want, and they're not digesting it well," she said. Parrish noted that fecal pellets are also common in modern organic-rich layers that fit the profile for future conversion into kerogen.

Hotbeds of productivity, such as coastal areas of upwelling, often feature the types of deposits associated with oil and gas, as Parrish's earlier research has shown. Upwellings of nutrient-rich waters from far below the surface often occur where currents meet continents, under the right wind conditions. Like rivers, upwelling brings nutrients to the algae and other marine plants floating near the nutrient-scavenged surface. These nutrients rejuvenate the many plants confined to living within the range of light at the surface. Where upwelling is a recurring feature on the seascape, productivity can become a regular enough theme that organic material accumulates below. This depends on location, though. Where winds promote upwelling in mid-ocean, most organic matter will be consumed or dissolved before reaching the seafloor. Around the coast, the shallower waters and greater supply of nutrients rushing in from rivers increase the chances that organic matter will make it to the bottom for potential preservation. Whether or not it's an essential condition, lack of oxygen can certainly help the cause in relatively shallow waters by encouraging the preservation of more organic matter. The burial (and potential preservation) increases during events such as hurricanes and major floods, when the amount of organic matter reaching the coast surges even as the number of creatures that might eat it falls.

Widespread deposits of marine organic matter occurred throughout the world's oceans during warm periods such as the Eocene and Cretaceous. Oceanographers generally have viewed the widespread deposits as evidence of bursts of productivity. Geologists, meanwhile, speak in terms of preservation, suggesting that a large-scale "anoxic event" robbed the ocean of oxygen that might have converted these deposits of organic carbon back into carbon dioxide. Either way, warm oceans have been linked to the creation of organic-rich deposits that eventually transformed into oil or gas. To turn kerogen into oil it takes millions of years of pressure-cooking, at temperatures of about 180 degrees Fahrenheit and the pressure that comes with burial at least 7,500 feet below the surface. If the concoction reaches 18,000 feet below the surface, it can crack into natural gas, or methane. While most hydrocarbons stretch on for dozens of lengths of the carbon chain, methane represents just one carbon atom with four hydrogen atoms framing it. A relatively pure petroleum product, it produces less pollution than oil and far less than coal. But while coal and oil generally stay put unless burned by a precocious hominid, natural gas can escape back into the atmosphere without assistance.

The Eocene appears to have started with a bang of methane gas release, and there's some evidence for methane bursts in the Cretaceous as well. But both hothouses also featured events that pulled carbon in the opposite direction. Using a popular example from about 94 million years ago during the mid-Cretaceous, a measurable spike in "heavy" carbon shows up in contemporary chalks, as Bradley Sageman and David Hollander note. This signature traces back to plant life. Always looking for the best molecules to build their structures, plants tend to reject heavier carbon when other options exist—which they did in the high carbon dioxide atmosphere of the Cretaceous. During a large-scale bloom roughly 94 million years ago, the ocean registered the burial of so much light carbon with these once-finicky, now-dead algae that the carbon remaining for chalk construction was decidedly heavy. Hence, the heavy-carbon spike in the carbonates actually reflected the widespread burial of intact algae—just the kind of event likely to lead to oil formation. Note that the timing appears to coincide with the large-scale burial of reefs around the world 94 million years ago.

❧

In time, with assistance from algae and other life-forms as well as the laws of physics, the planet put the lid on the high temperatures of the

Cretaceous, with its hothouse climate launched by extensive volcanic emissions of carbon dioxide (CO_2). As the authors of *The Cretaceous World* conclude, "What the additional volcanic carbon dioxide did was to accelerate feedback reactions which effectively increased weathering rates, added nutrients (along with dissolved CO_2) to the drowned continental shelves and also fertilized biomass production on land. If we want to look for the volcanic CO_2 of the mid-Cretaceous, it can be found in abundant Cretaceous coal and oil deposits and in huge Cretaceous calcium carbonate deposits."

The evolution of forams and coccoliths in the ocean, the expansion of flowering plants to all corners of the globe, and the return of the rain forests to the tropics all joined forces to pull carbon dioxide out of the mid-Cretaceous air. From a peak at the mid-Cretaceous 100 million years ago, carbon dioxide levels generally declined, albeit with various ups and downs along the way. Ancient temperature records similarly dropped from a mid-Cretaceous high. By 35 million years ago, ice sheets built up around the South Pole. By the late Miocene, about 7 million years ago, carbon dioxide reached the lows implied by the evolution of the low-carbon (C_4) plants. About 3.5 million years ago, ice began to build up around the North Pole. With the arrival of the Quaternary, officially designated in 2009 as starting 2.6 million years ago, the periodic buildup and contraction of the ice sheets had begun.

When the northern ice sheets contracted, wetlands and forests sprang up on the ground left exposed by the glacial shift. When the glaciers expanded, they plowed over the spruce forests and peatlands, while related climate changes around the globe converted many distant forests into grasslands. Coastal waters and their reef systems generally contracted when ice sheets grew and gained ground when glaciers melted.

Swamps and other wetlands create peat and coal, whereas coastal systems create an environment conducive to the creation of oil and gas. Meanwhile, forests collect carbon in their wood and soil. These responses lend a Gaian perspective to how forests and wetlands moderate climate over decades, millennia, and even longer. Some of the other Gaian services they provide, however, operate at much shorter time scales. Some of these services, including local cooling, cleaning, and climate control, operate at the scale of hours, days, and weeks. As the next chapter discusses, these instantaneous benefits help explain why the expansion of natural systems is so key during warm times.

8

Systematic Healing

More Ways Trees and Wetlands Boost Planetary Health

During an overnight foray into ancient redwood forest near Eureka, California, I felt as though I had entered another world. The high winds along the coast didn't penetrate into the tree-guarded sanctuary. The bird chatter seemed muffled by the plush carpet of pine needles where I rested, leaning my back against a cushy trunk that could have been a wall for all its size. I was lounging in the land of myth, home to the Bear, the Owl, the Salmon—and the ancient Redwood. Their tree rings confirmed that some of these trees had been here for more than 2,000 years, more than 26,000 moons. And unlike the longer-lived but much smaller bristlecone pines, redwoods continue to put on impressive volume throughout their existence. It would have taken at least half a dozen people linking outstretched arms to ring the individual bases of most trees in sight. Beyond the thick trunks, the upper branches teased outside my range of vision, even if I craned my neck. I felt as small as a squirrel in an Illinois forest of maple and oak.

A person can hardly help feeling insignificant among these giants that bring to mind J. R. R. Tolkien's Ents, the lumbering talking trees in *The Lord of the Rings*. Perhaps that helps explain why we've cut them down to the size of stumps across so much of their homeland. Since the mid-nineteenth-century Gold Rush, when some of our European ancestors invaded California in droves, we've shorn all but a few groves of coast redwoods (*Sequoia sempervirens*) and their equally formidable relatives, the sequoias (*Sequoiadendron giganteum*). Planting homes and highways in their place, old-growth forests of redwoods and sequoias survive on only about 5 percent of the Pacific Northwest land they dominated at the turn of the nineteenth century. Even then, their realm was a mere fragment of what it had been during warmer climes. The timing of their loss is especially unfortunate given the many Gaian services offered by these systems.

Redwoods illustrate the heights to which trees can go in serving as carbon collectors. In a world without humans, redwoods, swamps, and other forests would be expanding their range even now, as the warming planet called forth their services in clearing the air of some of its excessive carbon dioxide. But forests and wetlands do more than store carbon from formerly airborne carbon dioxide. The coastal redwoods also symbolize how forests and wetlands provide many real-time services that help moderate the impacts of climate change—including higher temperatures, greater risk of drought, and more extreme floods and winds.

When it comes to climate, exactly what we can expect of our future remains maddeningly imprecise. We know it's going to get hotter overall. But many of the important details remain unresolved, such as how this will affect local precipitation patterns, whether we will switch into a climate with an ice-free North Pole, and how these changes will affect food and water supplies for the world's growing population. We can expect stronger hurricanes and floods, but the where and when remain open to interpretation. We can expect more droughts, but we can't say how long ongoing droughts will last and where they will strike next. We can expect higher temperatures, but we can't nail down exactly when and where the heat will tip the scale into the danger zone for particular species, including humans in urban habitat.

Fortunately, we don't need to know all the answers to these questions to understand what we need to do to make our planet more resilient to the coming changes. There's no question that we'll need to make every effort to stabilize the global thermostat by cutting our use of fossil fuels like gas, coal, and oil. That's worth reiterating, but I won't dwell on it here because there are so many other fine books, articles, and Web sites that focus on that. This chapter continues this book's theme by considering how we can work with the planet's natural systems to moderate the warming and soften the severity of its impacts.

People are like the salt of Earth. We can complement the natural system—or do serious damage to it. Life-sustaining salts include calcium, iodine, and potassium as well as sodium and chloride. People flavor their foods with some of these essential salts, while herbivores seek out salt licks and predators supply themselves through the flesh of their prey. Yet too much of a good thing can be damaging to the health. Like too much salt, the sheer size of our population poses a risk to the planet's health. Our overwhelming presence on the planet multiplies the amount of effort needed to keep the system in balance.

When considering what to do, we must apply the primary dictum of health care to our planet: First, do no harm. We must stop destroying the systems that are working to keep the climate stable. Beyond that, we need to consider how to help the planet heal itself. Often all it takes is to remove the stress—which usually relates to some human activity. That's a lot trickier than it sounds in our current economy, where the bottom line ignores environmental costs and overlooks environmental services. In fact, the services provided to humans by wetlands and forests, including urban forests, go well beyond carbon counting. This chapter highlights the services that can help keep our planet livable in a changing climate.

❦

When I left the sanctuary of California's redwoods to sit in the bright sun on one of their long-dead brethren stretching along the coast, I met fierce winds, emboldened by the journey across the Pacific. I could barely hear my companion over their sweeping howl. Squinting at the sea through wind-whipped hair and glaring sun, I soon longed to return to the shelter of the redwoods. The climatic conditions in forests differ markedly from the outside environment. Under the forest canopy, whether the Ent-sized giants of the Northwest or the colorful maples and oaks of the Northeast, the air temperature almost never rises above our own body temperature. Whether among the thick-skinned ponderosas of the Southwest or the moss-dripping cedars of the Southeast, the air remains under 98 degrees in the shade. Trees cool the surface with their shade, creating their own environment beneath the boughs. The temperature difference is one of the most obvious effects of forests—although perhaps not to everybody.

A few years ago, a scientist suggested that Californians might stay cooler in coming years of global warming without the presence of their local forests. "While the carbon soaked up by California's forests reduces atmospheric carbon dioxide concentrations everywhere, cooling Crete, Cancún and Calcutta, the sunlight they absorb warms the state and the surrounding region. So, it might even cool us if we were to cut down those dark forests," wrote Ken Caldeira in a January 16, 2007, *New York Times* op-ed piece. I showed his column to my spouse, a non-scientist who works in adult education. "Has he ever been in a forest?" Bob asked incredulously. The drop in temperature under the shade of branches seems obvious to most people. Yet this scientist—who, in fairness, did note in his column that cutting down forests would be "nuts" given their many other valuable services—was not pulling his concepts out of thin air. One of several authors of a 2007 paper published in the *Proceedings of the National Academy of Sciences*, Caldeira was referring to their modeled estimates of how increased reflectivity of deforested areas would compare to the loss of the standing forests' carbon-collecting services. The authors, led by Govindswamy Bala, acknowledged that a global deforestation would increase the planet's carbon dioxide levels but suggested that deforested areas outside the tropics would be cooler locally anyway because they would reflect more solar energy than the tree crowns had.

The model had several flaws that marred its conclusions, though. Most glaringly, the authors overlooked the cooling effect of shade on the local environment. They employed a satellite view of the surface, so their assessment didn't penetrate into the forest canopy. Their definition of "local" temperature basically extended from the top of the trees to beyond the clouds—not the temperature that someone would register when walking on the shaded surface.

Call me superficial, but it's the surface I'm worried about. We animals and plants feel the heat at the surface. Could cutting down temperate trees or letting a deforested area remain barren actually help cool the planet, as Caldeira suggested? Not at the surface. Forests tend to look dark from space (as explained in chapter 3), so they reflect less incoming sunlight than lighter areas, such as glaciers or white sand. But reflectivity doesn't do much to cool the surface. Anyone who has danced across hot sand to get to the cool water or shaded grove knows this. Light-colored sand may well reflect more of the sun's rays than a forest, as a different climate modeler emphasized to me when extolling the reflectivity of North Africa's deserts. Yet despite all its reflectivity, the Sahara remains one of the hottest places on Earth. In unshaded deserts, daytime temperatures often soar beyond 120 degrees Fahrenheit. I'd prefer the Congo's rain forests any day—or night, as deserts soon turn cold. The averaging of cold nights with hot days can officially make annual temperatures lower in an arid desert than a humid forest, but that doesn't mean it's more comfortable. Or more conducive to life.

The reason global models can ignore the role of shade relates in part to how we measure temperature. Standard temperature measurements are made in the shade, typically in a covered box several feet off the ground. This reduces the wild swings in temperature that the sun's hide-and-seek behavior might cause on a partly cloudy day. But it also means there's no "glare factor" to account for the extra heat of direct exposure to sunlight. For example, University of Arizona researcher James Riley recalled being in Saudi Arabia in 1991 after the Iraqis had set many of nearby Kuwait's oil fields ablaze during the Gulf War. For months, smoke darkened the desert skies in the nearby Saudi city of Jubail. Many residents speculated that temperatures were running 20 to 35 degrees Fahrenheit below normal while the fires burned, he and his colleagues noted in a 1992 *International Journal of Biometeorology* article. In a follow-up e-mail exchange, Riley explained, "Basically, it was the shadow from the smoke that made us seem cooler, just as it would if you stood in the shade of a tree." Yet, as their paper documents, the shaded local thermometers sensed no real change in air temperature.

Few scientific studies capture the difference between surface heating under shaded ground versus bare ground, so it's tough to include this factor in a global-scale model. The local studies that have measured this factor, though, show a cooling effect of some 20 to 30 degrees Fahrenheit or more on a hot summer afternoon. For instance, F. Gomez and colleagues found that an unshaded temperature monitor warmed up to about 104 degrees Fahrenheit in the midday sun of Valencia, Spain, whereas a shaded monitor at the same site registered below 80 degrees. Under the midday summer sun in Phoenix, even the surface of shaded grass reached 104 degrees Fahrenheit—but the unshaded surface of nearby asphalt reached a scorching 140 degrees, Erin Mueller and Thomas Day reported in the *International Journal of Biometeorology*.

Both of the values cited here reflect the differences during the hottest part of the hottest season. That's when the difference tends to be the greatest, judging from those studies and a yearlong project carried out in piñon-juniper stands in New Mexico by University of Arizona ecologist David Breshears and colleagues. Burying their monitors roughly an inch below the surface, they found that peak temperatures of soil exposed to direct sunlight reached nearly 20 degrees Fahrenheit warmer than soils under shade on a July afternoon. The difference dropped during cooler times, and even went in the opposite direction in winter. For instance, January soil temperatures consistently ranked a few degrees higher under the trees, despite the shading factor, compared to the soil exposed to open air.

From a Gaian perspective, it's interesting that tree cover keeps the surface cooler when it's hot and warmer when it's cold. That's why it's more accurate to describe forests as "moderating" temperatures than as "cooling" them, even though the effect averaged across a year was to cool the soil surface of these piñon-juniper stands. The tree-inspired temperature moderation also influences how much moisture is available to plants. In the Southwest study, canopy cover kept soils above freezing some winter nights, unlike soils lacking protective forest cover. Canopy cover can help even during summer, it turns out. When Breshears and his colleagues extrapolated their temperature results to soil moisture, they concluded that shade probably would help keep soil surface more moist by slowing down evaporative water loss. That's because direct sunlight speeds up evaporation of soil moisture. Other researchers found that when the sun goes down, trees release into the soil some of the water they were holding. The extra moisture helps temper the next day's heating—another way forests moderate temperatures. The more water available for evaporation, the more energy can be used for evaporative cooling rather than surface heating.

For similar reasons, daytime air temperatures actually run slightly lower in parts of metropolitan Phoenix than in nearby desert, as a 2000 paper by Anthony Brazel and his colleagues showed. After initial puzzlement, the Arizona State University researchers who documented this realized that the water devoted to landscaping, "swamp" coolers, and backyard pools was evaporatively cooling these areas during the day, as a research colleague of Brazel's, Joseph Zhender, reported in a 2005 talk. The evaporative cooling of trees and other vegetation typically reduces air temperatures by some 5 to 10 degrees Fahrenheit on top of any surface cooling provided by shade. As when comparing the Congo to the Sahara, the landscaped city was hotter than nearby desert once the sun went down. At night, buildings, trees, and walls block breezes while radiating heat. In Tempe, the 10-degree-Fahrenheit average increase in urban temperatures since the 1970s comes from nighttime temperatures 20 degrees Fahrenheit higher averaged with stable daytime temperatures. What's more, evaporative cooling and shade from urban forests actually make the biggest difference during the heat waves destined to become more common as the globe warms.

❦

During a brutal July 1995 heat wave in Chicago, two of my brothers-in-law suffered heat illnesses after working outdoors in the searing humidity. Mark had spent hours outside doing survey work for a new development, while Jeff had been working with a team on a roofing job. Healthy men around thirty years old, they eventually recovered. But many did not. In time, researchers linked the weeklong heat wave to more than seven hundred deaths.

Heeding the wake-up call of this deadly heat wave, city officials have taken action to protect their residents from escalating summer temperatures. Led by Mayor Richard M. Daley, Chicago has launched a program to put "green roofs" on many city buildings, including City Hall. As of 2001, City Hall's roof features some 150 different plant species covering 20,000 square feet. Officials report that City Hall's rooftop temperature is running about 7 degrees Fahrenheit lower, on average, than neighboring roofs. As usual with vegetative cooling, the biggest difference shows up on hot summer afternoons, with the green roof up to 30 degrees cooler than its more exposed neighbors. Again, these plants also moderate temperatures by keeping winter temperatures warmer. The layer of soil helps insulate the buildings during cold Chicago winters, Green Projects administrator Michael Berkshire indicated on a city Web site. By 2007, building owners in the city had finished or designed 400 other green roofs totaling 90 acres.

Daley's efforts to green the Windy City received support, too, from a local study showing that urban landscaping apparently can help fight crime. In a 2001 paper, Frances Kuo and William Sullivan compared crime rates for ninety-eight apartment buildings in Chicago's Ida B. Wells housing complex, then one of the nation's twelve poorest neighborhoods. Residents had been randomly assigned housing, some into buildings surrounded by trees, others into comparable buildings amid barren pavement. Using police reports for two years, Kuo and Sullivan found that residents in the buildings with landscaping reported roughly half as many violent crimes, such as muggings and murders, as their neighbors who were surrounded by concrete and asphalt. Property crimes, such as thefts and burglaries, also were lower by nearly half. The study results not only refuted concerns that shrubbery could increase local crime but suggested that the value of urban forests goes beyond providing shade and evaporatively cooling the area. Although the researchers could not prove cause and effect in a non-experimental study like this, they found no other differences that could explain the reduced levels of crime rate better than the increased plant cover. Vacancy rate, the number of occupied apartments per building, building height—none of these factors negated the apparent influence of vegetation. No matter how they sliced it, the presence of plants remained an important explanatory factor for why the crime rate was lower.

The cooling effect of greenery, meanwhile, can become increasingly important as the climate warms. Air temperatures differed by 10 degrees Fahrenheit among Phoenix-area neighborhoods in two studies by Sharon Harlan and G. Darrel Jenerette, each alternating as lead author, with other researchers. The amount of vegetation accounted for about two-thirds of the differences in temperature in their study, Harlan noted during a 2008 talk at a City of Tucson urban heat island symposium. For example, a tree-lined, ritzy neighborhood remained some 10 degrees Fahrenheit cooler at times than a mostly barren area where immigrants from Mexico clustered together in high-density housing. "We're talking about microclimate as an environmental justice issue," Harlan said. "And as things become hotter, it becomes more important."

In fact, the temperature difference between low-income and high-income neighborhoods magnified during a Phoenix heat wave in July 2005, when air temperatures soared above 112 degrees Fahrenheit for four days in a row. "During the heat wave period, the difference between the two neighborhoods actually increased," Harlan said. The Phoenix neighborhoods with less vegetation spent more time in the "danger" zone. This had health implications. In a follow-up regional survey, the researchers learned that about a quarter of the residents in the neighborhoods with the highest heat exposure reported that someone in their household had suffered a heat-related illness during the summer. The higher outdoor temperatures made it that much harder to keep things cool indoors, especially for those without air conditioning.

As neighborhood income went up, landscaping generally increased—with some exceptions, including a new middle-income development. A new tract development located on the urban fringe, it should have been insulated from any inner-city density effect. Yet the researchers found that its heat load equaled and sometimes even surpassed that of the poorer, more densely populated neighborhood in their study. It turned out that the housing association of the new development had dictated that residents use Xeriscaping practices—that is, relatively sparse desert-adapted plants, often interspersed with gravel to hold down dust. The gravel acts much like asphalt when it comes to heating up the surface, as other research had found. By conserving water, Xeriscaping removes a source of evaporative cooling by plants. The heat load in this development and many others throughout Arizona also increases when bigger homes crowd onto an area. Standing next to the outside wall of a false-stucco Phoenix home on a summer evening can feel like sitting in front of a waning campfire. The more square feet of building material emitting heat per lot, the longer the slow burn continues through the night.

More cities will need to follow Chicago's lead and budget some money—and water, almost the same thing in some desert cities—to cool low-income, barren neighborhoods with trees and plants. As with most ecologically oriented solutions, the benefits of urban greenery extend far beyond cutting heat. Whether on the ground or on rooftops, plants provide habitat for birds and other wildlife. They insulate against noise and apparently even fight crime. Gardens and trees offer recreational opportunities, at the street level and when residents have access to the green roofs. They reduce air pollution, including the carbon dioxide floating around warming the environment. The soil hosting plants holds on to some of an area's rainfall, thus reducing the runoff that can flood streets. Soil and plants release the water in subsequent days and weeks, using some of it to evaporatively cool the air long after a storm has passed. Even grass can provide this service, if it has enough water to sustain its cooling operations. Trees, however, seem to have a talent that takes this water-based cooling a step further. Under certain conditions, forests can actually boost rainfall rates.

❧

Forests grow where moisture levels reach high enough to allow it, as keen observers have long noticed. In recent decades, though, researchers have realized that forests and the trees they contain actually help to generate some of the rainfall they need. This seems like a textbook example of how ecosystems can act in a Gaian way to improve the conditions for their survival.

In redwood forests, ways of promoting moisture include bringing airborne water as fog down to earth where it can be put to use. Researchers found that the amount of moisture staying on-site in California's redwood forests dropped by about a third in the absence of trees. Julia Butterfly Hill saw this process in action while perched high in the canopy of a 200-foot-tall redwood she called Luna. In her 2000 book, *The Legacy of Luna*, she provided an anecdotal account of her observations on one foggy day in 1997:

> The magical world of Luna is just phenomenal, down to how these trees disperse the water that falls from the sky. I was sitting in the fog one day, unable to see past Luna's branches, when I noticed that the needles at the top part of the tree are knobbier than the needles

lower down. Up high they look like gnarled fingers raking in the moisture from the fog and the rain. The water, drip by drip, gathers until it starts swirling down the trunk to the ground, over the smooth bark at the tip of the tree towards the increasingly shaggy bark down below, which absorbs more and more of the meandering flow. Toward the bottom of the tree, the needles become flatter and smoother. I imagine that's because they don't need to gather as much moisture. Instead, they act like a sprinkler system for the forest floor.

Along with collecting fog, forests promote rainfall. For one thing, they help keep water circulating through their own process of transpiration, a word describing how they evaporate water through their leaves. By putting water back into the air rather than merely letting it flow across the landscape into rivers, they help provide the fodder that forms clouds. This recycling skill reaches its heights in the Amazon. Global climate models typically fall far short by about a third when it comes to simulating the amount of rainfall observed in these tropical rain forests. Basically, the models have been unable to reproduce the extensive recycling that occurs as rain that falls on coastal forests is resurrected through transpiration to fall on forests farther inland, and so on down the line. Scientists have reached a growing consensus that forests increase rainfall, with the chemistry of rainfall and comparisons between intact forests and nearby agricultural plots providing convincing evidence.

Dense forests may also be generating rainfall in part because conditions under their canopy differ so much from the outside environment. That's a new theory by theoretical physicists Anastassia Makarieva and Victor Gorshkov and their colleague Bai-Lian Li. These scientists paired the logic of math with observations of climatic conditions beneath the canopy to suggest that forests create an internal temperature inversion that helps them retain moisture. In the Amazon rain forests, for example, daytime temperature peaks about 60 feet high in the canopy. That's where tall trees are intercepting sunlight and heat. Air temperatures below the canopy run about 10 degrees Fahrenheit cooler, thanks to shading and evaporative cooling. This creates an "inversion," with the canopy-level hot spot trapping air below. The "inversion" label comes in because it's the opposite of the atmosphere's usual pattern of having the highest temperature at the surface, with heat rising through air that cools with height. Hot air rises, but only as long as the air above it is cooler. A ceiling of hot air 60 feet high in the canopy acts as a lid.

This so-called temperature inversion may be more familiar as a phenomenon over cities, where tall buildings warming the air can create stifling conditions below. Urban inversions often lead to a buildup of polluting gases, with car exhaust unable to make a getaway into the upper atmosphere. An inversion in the rain forest apparently also serves to trap gases—namely, water vapor. When the air cools at night, the water vapor condenses back into liquid form. This can happen outside of the forest, too. The morning dew found in some places owes its presence to a similar situation, when night air cools down below its saturation point. In the forest, the trees help to create this condition more often, Makarieva and her colleagues suggest. The result: Forests keep their soils moist for the next day's work.

In cooler regions, canopy cover in forests and wetlands also could help keep water available through more of the year. In the southwestern study by David Breshears and colleagues described earlier, forest cover kept soil surface temperatures above freezing on many winter nights. In some cases, this could determine whether water is available. A tree locked up in ice is like a person going thirsty in the middle of the ocean—the water is not in a form they can use. And ice doesn't do much to ramp up the water cycle. (On occasion, it can sublimate directly into gaseous water vapor, bypassing the liquid stage, but it's mainly liquid water that evaporates.) Liquid water can help moderate temperature as well as boost precipitation, but frozen water just keeps things chill. So, once again, canopy cover can make conditions below more conducive to life. Wetlands, meanwhile, release methane, a compound that eventually transforms into other compounds including water (described in chapter 6). Granted, the amount of methane produced by wetlands doesn't add up to much water proportionately; water is the world's most abundant greenhouse gas. Still, there's no denying that trees need water to grow. Extensive tree cover, whether in forest or swamp, has another fringe benefit: It helps block the wind. Forests and wetlands even slow hurricane winds.

❧

Anyone who has spent a winter in the Windy City can vouch for the important influence of wind on temperature. The so-called wind chill factor can turn a cold day into a miserable one. Personally, I always thought that DeKalb, Illinois, topped even Chicago for bone-cutting bluster. Maybe it was just that I spent much of my time in Chicagoland on streets named "Maple" and "Greenwood." In DeKalb, I had to walk across an open field to get from my dorm to my classes—ten minutes of tempestuous torture. So one of the ways forests moderate temperature is by moderating winds. February days in DeKalb notwithstanding, this role becomes even more important in a warming climate that spurs on more intense storms.

Forests decimate normal winds. Wind speeds within a mixed oak forest in Tennessee in January had dropped by 88 percent of those outside the forest. In European forests, they slowed by 93 percent within a few hundred feet of forest boundaries. Most of the decline occurred within the first hundred feet. Similarly, winds confronting tropical forests in Brazil slowed by 80 percent within a few hundred feet of entering tree territory and dropped to negligible speeds in about half a mile. The long-appreciated effect forests have in blocking the wind has led farmers around the world to create wind breaks from trees to protect their crops, and the soil supporting them.

Forests can temper even hurricane winds in time. A storm's smooth sailing over the ocean hits a rough spot when it encounters the bumpy canopy of a forest. Tropical cyclones rarely maintain hurricane strength for more than 100 miles after hitting land. Lack of fuel slows them down, but so does friction. And tree canopies provide some of the most friction. Trees of different heights force winds to waste energy navigating and crossing paths. Storms also expend some of their force in knocking trees around. Hurricane Hugo in Puerto Rico flung large trees around for hours. Of course, some trees became completely uprooted or snapped under the pressure. While standing, though, they contributed wind resistance. Those surviving the battering did so throughout the storm.

"Trees slow down the wind and they don't provide heat," hurricane expert Kerry Emanuel told me during a lunchtime conversation at a 2006 conference in Palisades, New York, on tropical cyclones and climate. Between bites of his sandwich, he explained why forested wetlands—that is, swamps—do a better job of slowing hurricanes than open water. "There's a certain amount of heat stored in hot swamp waters. There's just not enough fuel in a swamp," he said, following the habit of those in his field of regarding warm waters as "fuel" for hurricanes. "There's fifty meters of fuel in an ocean, where you'll be lucky to get one meter in a swamp." A meter, roughly a yard, would come about waist high on a tall adult. Warm waters about 30 feet deep, typical of those near the coast, can really empower hurricanes, Emanuel noted, because there's no underlying layer of cool water to slow them down or land close to the surface to limit the supply of fuel and provide friction.

Mangroves and marshes help keep land from turning into open water. Wetlands stable enough to form land out of roots and debris really help put a brake on a hurricane's destructive power. Where they haven't been starved of supporting sediment, barrier islands also serve to slow hurricanes approaching the coast. There is nothing like friction to slow down a moving vehicle, whether it's carrying cargo across the country or transporting a heat load out of the tropics. Any extra soil between a hurricane's landing and its arrival at civilization's doorstep helps, much as a longer runway gives an airplane more opportunity to slow down.

What's more, wetlands absorb storm surge and slow flooding rivers. Soggy as they are, they will sponge up storm water flowing over them. Mile by mile, they lower the risk to homes. They also slow floodwaters from rivers. Like skateboarders, rivers need some slant to gain real speed. When they hit the flat areas where wetlands thrive, streams slow down. As their velocity drops, so does some of the sediment they worked so hard to carry. Thus wetlands help maintain and build land.

These days, the forested wetlands known as swamps often occupy only a sliver of land, so their protective powers are limited. In New Orleans, for instance, urban development put a straitjacket on the Mississippi, rushing its flow to the Gulf by removing curves and lining straightened channels with concrete. Before that, the river had meandered through the area across an intestinal-like winding path of some 100 miles, nourishing wetlands such as cedar swamps around almost every bend. Since Katrina, New Orleans officials are reevaluating wetlands as a means of protecting their city from the winds and rains that come with major storms. There's even talk of diverting the Mississippi outwash from its current location—where it provides the sediment that's building the "bird's foot" of land off the delta—closer to the city so it will build wetlands in a more strategic location in relation to New Orleans. This idea has merit, given the land-building potential of sediment and the wetland vegetation likely to colonize suitable sites. Before diverting the direction of the Mississippi plume, though, officials would do well to consider the quality of the water it contains. The plume feeds a huge dead zone that crops up every summer as algae crowd around to enjoy the fertilizers coming down the pipeline (as described in chapter 7). Before directing that movable feast inland toward productive coastal waters, planners might want to consider a couple of other related roles performed by forests and wetlands, and even gardens: They absorb storm overflow and improve water quality.

❦

"Here's where the rivers are today," Brad Lancaster told a group gathered at the University of Arizona's Water Resources Research Center. On the screen was an image of Tucson streets flowing with knee-deep water, much like the view from my porch of a street I affectionately thought of as the River Mabel during monsoon season. "We're truly desertifying our so-called desert," he said, noting that we rush water off the land before it can be absorbed. Lancaster teaches residents of Tucson and others how to harvest that water for growth and has written two volumes on *Rainwater Harvesting for Drylands and Beyond.* During a tour of his Tucson backyard, he showed how this way of thinking could bring the landscape to life. He had contoured it so that water would collect in the vegetable garden, dug a little deeper than the surrounding soil. He funneled water from his roof onto the land and into a 1,200-gallon cistern, available for watering his garden during dry spells. Nearby orange and fig trees flourished from water draining from his zinc-roofed workshop and his outdoor washing machine and shower. Added to the mix was other "graywater" recycled from inside his home (sanitary water from everything but the kitchen sink and toilets). In front of his property, he had cut a gap in the curb to usher in water. Any street flow would nourish the mesquite and other native plants thriving in the depressions he had created on either side of the footpath meandering past his home.

Efforts by Lancaster and others have helped launch a sustained interest in using roof overflow and household graywater to create backyard oases. On a lot where the home takes up about half the area, collecting rooftop water can amount to doubling rainfall. If an inch of rain falls, that means if all the water stayed on-site it would total an inch deep at any one spot. So if 1 inch of rainfall is funneled from the roof onto the vegetated half of the lot, that half effectively receives 2 inches of rainfall. In an average Tucson year of around 12 inches of rainfall, then, roughly 24 inches would reach the landscape or cistern (although a little is lost to rooftop evaporation). This puts it into the range that can support species besides native mesquite and paloverde (which naturally tend to grow in depressions or valleys that collect water) into the realm of pine and oak. With further concentration, such as by collecting water off paved streets or driveways, the water could sustain fruit trees, even citrus. Letting trees grow equates to putting sunscreen over the land. Trees and soil also absorb storm water, thus keeping it on the landscape rather than flowing down the street.

A lack of wetlands and an overabundance of pavement clearly affect the quantity of water flowing down Tucson streets or pooling in New Orleans. These factors also affect the quality of the water—especially during floods. Hurricane Katrina brought this message home to the New Orleans residents who had to trudge through its floodwaters. Dawn Griffin, a longtime New Orleans resident who managed to maintain both a sense of humor and hope despite her harrowing experience a few months earlier, described her brush with Katrina floodwaters over a beer at the French Quarter's House of Blues. For a few days after the storm, the fire company stationed at the Grandstand Fairgrounds across from her house kept an eye out for her. But when they began preparing to leave, they convinced her to seek out the promised helicopter rides. "In order to get there I had to go through neck-deep water," recounted Griffin, who stands a few inches over five feet tall. Downed trees acted as submerged obstacles. "People asked, 'Can you swim?' That's not the point. This water was not like being in a pool. The stench was unbelievable. There's nothing to describe that smell."

Katrina floodwaters in New Orleans represent the worst-case scenario of a phenomenon that occurs to varying degrees in any city flood. While the wetlands they often cover help purify water, cities magnify water pollution. Like farms, cities add waste, fertilizers, and pesticides to the water. Floodwaters can carry sewage from septic tanks, storm overflow, and sludge fields. Along with lawn-care products, they can sweep up remnants of industrial-sized vats of chemicals from factories. Petroleum products from gas-pump overtopping and backyard oil changes add to the brew. Even when storm waters don't rise enough to cause destructive flooding, the quality of water reaching rivers can suffer. The asphalt and pavement coating many cities can funnel the toxic mix straight into the river. If its banks are lined with concrete as well, there goes that one last chance at running the water through a carbon filter of soil.

Water purification occurs on the landscape everywhere wetlands exist. The cleansing power of wetlands is so well known that it has even been tapped for sewage treatment. A laboratory situation really brings home the message that plants can thrive on the stuff we flush down our toilets. Ohio's Oberlin College has such a laboratory. Known as a Living Machine, it offers a glimpse of how wetland vegetation can clean water in real time. It's part of the Center for Environmental Studies, designed with sustainability in mind by visionary David Orr in conjunction with architects and students, as he describes in a 2006 book, *Design on the*

Edge. Orr graciously gave me a tour of the center on a July day in 2006. Looking more like a coach than a pioneer, with a red baseball cap over his short-cropped light hair, Orr pointed out some of the sustainability features. There's a two-story wall of windows facing the winter sun to cut seasonal heating costs, as well as a student-run garden. We admired a solar-paneled parking ramada that makes energy and shade at the same time.

Finally, we stopped in front of the Living Machine, a water-treatment plant designed by a fellow visionary, John Todd. I had wanted to see one in action ever since meeting Todd in the mid-1990s, when he came to Puerto Rico to share his ideas about plant-powered sewage treatment. Orr opened the glass-windowed door to reveal the ingenious interior. Out came a whiff of the moist soils of Louisiana or the lush tropical rain forest of Puerto Rico. It's a scent filled with life and the promise of its continuation, with no detectable undertone of sewage. Ferns dwelled within, alongside waxy-looking, large-leaved plants that speak of climates much warmer than Ohio. A monitoring system shows that the water treated here comes out cleaner than that from a typical sewage-treatment plant, Orr pointed out. Even so, protocol requires that it go to support outdoor ecosystems rather than indoor drinking fountains. He led me outside to show me the beneficiary, a pond in front of the building that features water lilies and resonating bullfrogs.

Perhaps it shouldn't be so surprising that plants thrive on our waste products. Clearly they appreciate the carbon dioxide we exhale as waste. Everything we're made of, including our bodies once we discard them at the end of our lives, looks like compost to plants. Plants, including microbes, love nitrogen in all its forms—even as ammonium, a compound one quick step away from urea, the main ingredient of human urine. To plants, urea is fertilizer. The Chinese have harnessed the fertilizing properties of urine for centuries, perhaps millennia. Farmers in the midwestern United States also know the value of waste products. I recall some memorable days in DeKalb, whose logo features a flying ear of corn, when the air carried the pungent scent of cow manure. In fact, the real problem with sewage is that plants love it too much. With all that nitrogen, clear-flowing rivers soon become algae-filled channels. Where rivers transporting sewage or other fertilizers enter the sea, the resulting algal blooms can create dead zones. The other problem with putting sewage in water is that it can transport disease to other people. From a human as well as Gaian perspective, then, it makes more sense to dispose of sewage on land, where it can nourish the kinds of plants we appreciate—such as cedar trees.

Sewage effluent from New Orleans waste-treatment plants could support cedar swamps, for example, as in a creative proposal by John Day Jr. of Louisiana State University. After all, at that stage of its treatment, effluent is mostly water—just the thing to construct wetlands. The extra nitrogen in the water also suits these ecosystems, which use nutrients from the water to grow. The waste-to-wetlands vision takes other forms as well. Midwestern wetlands could help bring the Gulf of Mexico's dead zone back to life, Day and lead author William Mitsch suggested in a 2006 paper. Wetlands purify water by collecting toxins and retaining nitrogen, an ingredient that promotes algal blooms and thus coastal dead zones. Restoring about 5.4 million acres of wetlands throughout the Mississippi basin could cut the amount of nitrogen fueling the dead zone by about 40 percent, they calculated. That's about twice as many acres of wetlands as were destroyed in the United States in one decade, from the mid-1970s to the mid-1980s, as Ralph Tiner reported in his book *In Search of Swampland*. But it's only a fraction of the 130 million acres that were obliterated since the 1700s, Tiner estimated.

The estimates on how much nitrogen wetlands can extract came from research on actual wetlands, including some that Mitsch and others constructed on the campus of Ohio State University. "I call them Chia wetlands. Just add water and they sprout," Mitsch said. He gave me a tour of these kidney-shaped wetlands on a sunny morning in August 2009. The wetlands now cover about 25 acres of the Ohio State campus. While Mitsch and others planted about twelve different marsh plants in one area, in another they followed a concept known as "self-design"—they just added water and let the wetland develop on its own. "Mother Nature is a chief contractor," he joked. "The one that we planted still has a little more diversity. But the one that Mother Nature so-called planted, for the longest time it had more of what I call 'power.'" For instance, its productivity rates soared, so it was converting more carbon dioxide into mass than its more domesticated counterpart. During the tour, I was especially impressed with the birds, including a leggy blue heron, and the fresh, almost flowery scent of these lush systems. It was difficult to reconcile the lush plant life with the images showing the construction of these wetlands from barren, bulldozed soil only fifteen years earlier.

Mitsch and Day have been touting the value of restoring wetlands in the Midwest for many years, but the recent push for ethanol made from corn has made it even more crucial. "The kind of money they can make for growing corn, not just for food but for energy, has made it very lucrative. They've got corn growing everywhere now," Mitsch said. This is creating problems for the environment as well as for the world's people. Many people already lack adequate food, and these efforts to make fuel out of food are exacerbating their hardships. The environment suffers because farms are designed to rush extra water off the crops, shifting fertilizer-laden waters into local rivers that feed into the Mississippi and, eventually, the Gulf of Mexico. "The drainage system in the Midwest rivals any city," Mitsch noted. "The drainage tiles under the landscape get the water out of there as quickly as possible." A diversion into wetlands would improve water quality from local rivers all the way down to the Gulf of Mexico.

❧

Where wetlands thrive, they collect water from throughout the landscape. Then they purify it. In dry times, they serve as a long-lasting source of moisture. In many areas, they provide havens for wildlife and a refuge from surrounding heat. Often, wetlands and estuaries serve as a mixing ground for species from throughout the landscape. Thus, wetlands can assist species in adapting to climate change—including at the genetic level.

Biological diversity takes three different forms, and wetlands and estuaries can increase all three of them. Diversity of species is the most familiar. Estuaries and wetlands serve as breeding grounds and nurseries for a variety of species. Shallow lakes and marshes are favored by birds, which cluster to the rivers in desert lands. In arid lands like Arizona, the San Pedro River and its associated wetlands register one of the highest biodiversity levels of species in the region, serving as the meeting ground for birds migrating south from Canada and those flying north from the tropics. In the United States, wetlands provide habitat for a variety of species, including a third of the birds, a sixth of the mammals, and half of the fish listed as threatened and endangered. In the humid tropics, wetlands have a somewhat lower biodiversity compared to the species richness of the rain forests that surround them. But that's just because it's tough to beat tropical rain forests for diversity. Perhaps because the species populating lowland rain forests never underwent burial by massive ice sheets, a tremendous variety of species abounds. The 35-mile by 100-mile tropical island of Puerto Rico has as many different tree species as the entire U.S. mainland, for instance.

Seen at a larger scale that encompasses both the wetlands and surrounding rain forest, though, wetlands expand regional biodiversity by nurturing different plants and creatures. This brings up a related type of biological diversity: ecosystem diversity. In Puerto Rico, palms thrive in wetlands, such as in upland valleys, while bamboo may fill the waterside role often played by cattails farther north. The variety they bring to the landscape can attract and sustain species that might not otherwise exist there. In southern Arizona, rivers and the cottonwoods and willows that fringe them make up a minute portion of the landscape, but they draw animals from miles around. These local watering holes, which some call riparian wetlands, provide a crucial service in dry lands. Almost everywhere forests exist, rivers and wetlands intermingle with them, creating a mosaic of habitats that sustains diversity among animals as well as plants. The mix of ecosystems adds a level of complexity to a region, creating more and varied niches for the local species.

Wetlands and estuaries also can prove key to maintaining genetic diversity, the final major type of biological diversity. A species may contain either a high or a low amount of diversity in its members' genes, which carry the traits passed on to future generations. When a population of animals or plants becomes isolated, genetic diversity can decline. This can be a real issue with endangered species, especially if only a few hundred individuals remain in existence. But a lack of genetic diversity could limit a species' capacity for adaptation almost anywhere species become cut off from others of their kind, whether on an island or in a watershed, as ecologist Ariel Lugo told me. This points to another value of estuaries—as meeting grounds for aquatic individuals otherwise isolated in watersheds that drain separately into the ocean. "Watersheds are parallel to each other. Their waters are isolated one from another. But they come together in the estuaries. They remix the species and keep nature on its toes," Lugo added. "This is very important for climate change. You need the mixing. You need to bring all of genetics to bear."

Wetlands and estuaries can even serve as refuges during local climate changes that disrupt surrounding forests or nearby coastal waters. With the additional carbon they export to the sea, they can dish up sanctuary from the heat. In a New Zealand reserve called Gaer Arm, decaying leaves from upstream forests stain rivers brown with tannins. A natural dye used in leatherwork, tannins actually act like sunscreen on the water. With this extra protection from the sun's penetrating rays, the water can shield a variety of creatures normally found in deeper—that is, cooler and less radiant—waters. In a way, the tannins "dim" the amount of sunlight penetrating the water, much as certain pollution particles can mingle with the clouds to block sunlight. This pollution effect, mainly from sulfur compounds like the ones that keep asthmatics inside on hot days in big cities, has shielded the planet from some of the impacts of global warming. In a similar way, tannins in the water give deep-water species a local place to survive rising water temperatures. If the world continues to warm, these refuges could give species time to adapt—or just provide a shelter so that some cooler-water species will be around to spread out during a future cooling.

❧

Wetlands form in low-lying areas where water pools and sediment gravitates. They collect creatures and materials from throughout the landscape, including debris, leaves, and other organic material. When wetlands remain intact, mangrove trees and marsh grasses trap incoming sediment and debris and use it to form terra firma. By creating land with flood-deposited rocks, logs, and piles of sediment, coastal wetlands work to reduce some of the negative impacts of sea-level rise. The skill of wetlands and peatlands in building land was covered in chapter 6, but it's worth remembering in the context of a global warming that can bring stronger storms and higher seas. Soils in wetlands and forests serve as carbon filters that purify freshwater. They also help store floodwaters, including storm surge from hurricanes. Wetlands and forests block winds, too, and provide friction that even slows hurricane winds over time. They evaporatively cool the landscape. Trees make the biggest difference on summer afternoons and during heat waves. They even help generate rainfall. Skills like these lend further insight from a Gaian perspective about why a warming planet produces a protective shield of wetlands and forests.

Ice ages typically came with lower greenhouse-gas levels and lower temperatures—and also lower seas, lower precipitation rates, and lower plant biomass. Added to that were lower weathering rates and lower chalk and reef production. Hothouse climates came with higher greenhouse-gas levels and higher temperatures, and also higher seas, higher precipitation rates, and higher plant biomass—not to mention higher rates of weathering and chalk and reef construction.

Now here's a puzzle: What on Earth happens when a newcomer to the

Gaian system interferes with the natural responses, honed over hundreds of millions of years, to a warming climate? What happens if the planet has higher greenhouse-gas levels, higher temperature, and higher precipitation rates—yet with the unusual combination of lower biomass, lower water tables, and lower quantities of weathering products reaching the sea because of extensive development, logging, groundwater pumping, and river diversion? That's the experiment we're running now.

9

Conclusion
What Would Gaia Do?

Wetlands have been called the planet's kidneys, given their role of purifying a dilute toxic stew of chemicals into potable water. This function only works in a healthy system, though. The ongoing destruction of wetlands leaves our planet with the equivalent of one kidney to handle the toxic load of a meth addict who subsists on French fries and whiskey. Forests have been called the planet's lungs. Again, our hacking away at these systems equates to asking an asthmatic chain smoker to run a marathon. Are we asking Gaia to do too much with too little? How can we help, instead of hinder, efforts by our planet and its natural systems to control climate and support life?

Gaia theory originator James Lovelock has compared humans to the planet's central nervous system, akin to Gaia's brain. It seems the warning signal about global warming has finally penetrated our thick skulls. Now let's hope we will feel an impulse to do something about it. Just because a system or a planet can heal from a major assault doesn't mean it will. The wrong behavior can produce a turn for the worse. A person with a fever can recover, under the right circumstances. But if that person tries to pretend nothing's wrong, smoking and drinking the nights away, she or he could turn a mere chest cold into fatal pneumonia. A similar concept applies to the planet. Anyone who imagines that we don't have to worry about our bad habits because Gaia will take care of the mess is missing the point. We *are* Gaia. We're the ones who must make repairs, along with the other living systems on the planet. "We need to begin to understand our symbiotic relationship to Earth," Justin Willie, a Diné of the Navajo Nation, reminded during a 2005 water summit held in Flagstaff, Arizona. "The problem is the solution. It's us."

That's not to say that we should attempt to *take over* the regulatory functions of the planet. In the Earth theater, many of Gaia's longtime regulators seem to have their roles down somewhat better than we do.

Sure, trees and wetlands pulled down a bit too much carbon when they first evolved, as suggested by the 80-million-year long ice age that spanned parts of the Carboniferous and Permian. But ecosystems have evolved refinements over the years. Although the mechanisms remain in outline form, Gaia has managed to avoid a fatal heating spell even during warm periods and to resist a permanent plunge into temperatures icy enough to stop life in its tracks. Humans have been a recognizable force on this planet for maybe 2 million years. Every day, ecologists are discovering more about the many subtle interactions among species and ecosystems, but there's still a lot to learn. And yet some people imagine that we could run things better than the complex natural ecosystems that have evolved over hundreds of millions of years.

Perhaps we're overrating our own species' ability to control climate. Here are a few of the proposals out there about how to fix global warming: Shield Earth's surface from the sun with giant screens floating in orbit. Add more pollution to the upper atmosphere to block out some of the sun's rays. Erect huge metal "trees" circulating sulfuric acid to collect airborne carbon dioxide. Like any techo-solution, all of these create new problems. Blocking sunshine would reduce the amount of energy available for photosynthesis. Adding more pollution would aggravate health conditions like asthma and heart trouble and ultimately make the ocean and other water bodies more acidic. And why use energy and mined metals to make "trees" that poison the air and any creatures who step foot into their acidic midst when Gaia can make trees that provide food and shelter, sunscreen and windbreaks, and flood-control and drought-prevention services even while collecting carbon dioxide and other pollutants? What's more, planting and maintaining real trees would create far more jobs than any techno-solution.

Desirable solutions will be recognizable not only by their ability to slow down global warming and reduce the impacts of its symptoms, but also by how they affect other aspects of the Earth system—including its people. As many know, the world's poor people carry a disproportionate share of the burden from global warming and its impacts. Bangladesh is one of the poster countries for global warming threats, given its millions of people eking out a living in low-lying lands subject to sea-level rise, hurricanes, and storm surge. Yet we have our own Bangladesh here in the United States, in the bowl of New Orleans. America's poor elsewhere, too, face challenges from the changes that come with climate shifts. Whether living on sinking land in Louisiana, in the concrete-slabbed inner city of

Phoenix or New York, or on tribal lands with boundaries that remain firmly in place even as species shift, the people most vulnerable to the coming changes are those without disposable income, insurance coverage, and second homes.

Along with America's poor, many members of the middle class are probably more susceptible than they realize, given insurance loopholes that cover hurricane winds but not floods, wildfires that can consume towns, and power outages that are most likely to strike when air conditioners are running full blast. Solutions should help alleviate as many risks as possible. Similarly, proposed solutions that make things worse at some scale or for some people must be acknowledged as having this effect. Going into the details of specific policy approaches will have to wait for future writings. In this concluding chapter, though, I would like to sketch a broad outline of some considerations that should be on the table when considering how a Gaian perspective of the values of natural systems could inform the complexities of social policy.

❧

Clearly, natural systems, especially forests and wetlands, are taking up carbon. Globally, this has been documented well enough over the past decade or two that the argument comes down to the details of exactly how much and where. Of course, these details matter. Given the proper attention to detail, though, there's value to putting money on the table to safeguard and expand these systems and their carbon cache. With the right approach, giving forests and wetlands credit for the carbon they collect can provide funds to help protect them. Recognizing the many other services they provide could add incentive and dollars for their protection. Their diverse portfolio of services includes flood control, drought prevention, wind reduction, habitat creation, water purification, sun protection, and temperature moderation, to name a few. Society needs to find a way to expressly support these systems, whether via carbon credits for forest and wetlands restoration, environmental equity to cool inner cities, or nature-based insurance for vulnerable coasts.

The ability of natural systems to draw carbon down from the air goes far beyond theoretical. Despite humanity's ongoing destruction of forests, globally these systems still manage to take in more carbon than they are releasing. They're taking up all the carbon released in the burning and razing of their brethren. Then they're coming up for air to take up even more—enough to account for at least a quarter of humanity's annual carbon emissions from fossil fuels. That's an amazing feat. Thanks to forests and the ocean, only about half the carbon dioxide from our fossil-fuel emissions stays in the atmosphere to warm the planet.

Given this help from Gaia's natural systems, our own goal of stabilizing carbon emissions seems more manageable. At least we know there's a mechanism for helping us to balance out some of these greenhouse-gas emissions. Taking the work by natural systems into the equation could even suggest a goal. If we cut the world's carbon emissions in half, we'd at least be putting them into a range that could be handled by natural systems, based on what we've seen this past half century. That's a bit simplistic, as some of their uptake might relate to having so much carbon dioxide in the air. But it's worth considering as a baseline goal. In fairness, every world citizen deserves an equal share in the carbon allotment. Given how much more carbon dioxide the average U.S. citizen emits compared to other world citizens, this approach would still require a typical American to reduce emissions, in effect, by some 75 or 80 percent.

However necessary, that's an ambitious goal. Given the extent of reductions needed by the typical American, it seems likely that carbon offsets will need to be part of the equation, at least in the short term. Even climate-change champion Al Gore has turned to carbon offsets as a way to counteract the greenhouse gases emitted from heating his home and jetting around to spread the word on global warming. Gore mentions the value of planting trees to offset carbon released from fossil-fuel burning in director Davis Guggenheim's documentary *An Inconvenient Truth* and his own book of the same name. So—does this really work? Or do such carbon credits amount to selling hot air, as some critics charge? The answer depends on the approach taken.

The fact that much of the carbon retained by trees is stored in wood makes it feasible to measure the actual carbon storage fairly accurately. There would have to be a system in place to measure real, aboveground growth over a reasonable time frame. Between on-the-ground measurements of growth gains, remote sensing using satellites, and even the collection of tree-ring cores in temperate areas, it should be possible to reasonably account for how much carbon is coming down into wood.

The concept of giving credit for soil carbon collected also is rooted in reality, although the difficulty in measuring the variable carbon uptake beneath the surface poses extra challenges. Wetlands also can be tricky to assess because some of them release methane. It's expensive to measure methane and carbon releases, and it is no easy task to estimate how much carbon is being collected in soils, including the soil in wetlands. The global increase in methane levels cannot be blamed on wetlands—wetlands have lost ground nationally and globally even as methane levels have escalated. Still, the potency of methane as a greenhouse gas leaves some people wary of systems that emit any at all. At a minimum, it's worth remembering that mangroves and other coastal wetlands emit almost no methane even while collecting unusually high amounts of carbon. Some subtropical freshwater swamps, too, such as those in Florida and Georgia, have such a high ratio of carbon uptake to methane release that the latter could be considered negligible.

While it's valid to credit forests and other natural systems for their carbon-collecting talent, it's all too easy for this valid concept to be stretched beyond the reasonable. For instance, current Internet "sales" are often based on the concept that planting one tree will offset a (metric) ton of carbon, the equivalent of 3.6 tons of carbon dioxide. This gives the tree too much credit. The nonprofit conservation organization American Forests estimates that the average tree captures about 1 ton of carbon during 40 years of growth. By extension, then, it would take about 40 trees to offset a ton of carbon in the same year it's emitted. And we're not talking seedlings here, either. It would probably take a seedling at least a decade to grow large enough to carry its perceived weight in carbon emissions. Rounding off and switching from carbon to carbon dioxide, a typical American would need to maintain roughly 75 trees to reduce by 75 percent his or her annual carbon dioxide emissions from personal habits such as driving, flying, lighting, and heating use. That's per person. And it might take planting three times that many to get enough trees to survive the seedling stage, depending on conditions.

The good news is that every year those trees continued to thrive they would offset a similar amount of emissions. In fact, their ability to withdraw carbon dioxide from the air increases as they grow, well beyond 40 years for almost all but the most pioneering species (for instance, 30 years may be a typical life span of the weedy *Cecropia*, the big-leaved tree known in Puerto Rico as yagrumo). Many trees continue taking up more carbon dioxide than they respire throughout long lives, contrary to popular belief that "vigorous" growth occurs only in the tree's youth.

This exercise suggests that those who are serious about offsetting their carbon emissions might consider thinking in terms of buying land with forests or wetlands to protect or planting trees on lands that could support them. This could be done as individuals, working with others, or perhaps as part of a conservation easement offered by reputable environmental groups. By having specific parcels of land identified in public records indicating who supports it, the purchaser would be assured that the trees planted there are indeed thriving as promised. Tree-planting efforts should also include some acknowledgment that fires happen. Perhaps groups interested in selling carbon offsets should be restricted to annual sales, so the tallies could reflect reality (including loss to fires), rather than good intentions, when it comes to carbon counting.

❧

The increasingly common runaway wildfires plaguing the U.S. West represent in-your-face evidence that some of the carbon stored in forests around the world could go up in smoke. Wildfires make only a small dent in carbon stores, though, when compared to growth occurring in forests. In a few less populated states, namely Alaska and Idaho, greenhouse-gas emissions from fires can surpass societal emissions in some big-fire years. This finding by researchers Christine Wiedinmyer and Jason Neff made headlines in November 2007. Fires from U.S. forests can contribute 10 to 15 percent of annual greenhouse-gas emissions from global forests, based on Wiedinmyer and Neff's estimates for U.S. forests and other estimates for global emissions they cite.

As with any exercise that involves putting numbers on carbon-exchange processes, there's a range of error inherent in estimating fire emissions. Wiedinmyer and Neff note that their estimates of fire emissions are two to five times higher than one produced by a version of the Global Fire Emissions Database. Meanwhile, their estimates of greenhouse-gas releases from Oregon's Biscuit Fire of 2002—which burned half a million acres of forest, roughly comparable to Arizona's Rodeo-Chediski fire that same year—run about three times higher than an estimate made by B. E. Law and colleagues using a different approach. In this case and many others, the scientists used "carbon dioxide equivalent" to compare emissions. This allows emissions from methane—a major component of wildfire releases—and other gases to be compared, taking their different potencies into consideration. Emissions estimates require making assumptions about burn severity, the proportion of trees versus grasses, differences in wood density and amount of deadwood across the landscape, and the types of emissions spewed out when they burn. What's more, different sections of the forest burn to varying degrees. As

Wiedinmyer and Neff note, there's no method in sight for reducing the uncertainty in these large-scale assessments. Too many variables prevent researchers from honing in on a precise number.

But, just as global forests manage to consume at least the amount of extra carbon dioxide emitted by forest fires in an average year (chapter 5), U.S. forests soon retrieve the greenhouse gases emitted by wildfires. The nation's forests consumed twenty-five to thirty times more carbon dioxide equivalent than they released in fires, based on Wiedinmyer and Neff's comparisons with five-year satellite estimates of net primary productivity. The years they analyzed, 2002 to 2006, encompassed a large-scale drought that spurred on widespread wildfires. If anything, the ratio of carbon collected compared to carbon released by forests would go even higher during wetter periods.

Still, fires do throw a curveball into the concept of protecting or replanting forests to reduce carbon dioxide levels. What if some of that investment goes up in smoke? That's one reason those who can afford it—and sometimes those who can't afford to lose it—have insurance policies. Insurance policies acknowledge that accidents happen. Winds wail. Rivers swell. Coastlines advance, and in the future may not retreat to their current shores. And fires blaze away. There is no suppressing fire over the long term. A century of stamping it out has only turned U.S. forests into bigger firetraps. Perhaps by building in the concept of insurance—including treatments that help ensure against stand-destroying blazes—fires could be accepted as the natural phenomenon they are in the context of carbon-offset projects. Official approaches ignore carbon releases from wildfires, a sleight of hand that has the effect of devaluing efforts to restore some western forests to more fire-resistant conditions.

❦

Many pine forests in Arizona's White Mountains contain too many small trees that could torch neighboring old-growth trees and homes. Preventing fires, as well intentioned as it is and has been, allowed the growth of saplings that would not have survived a natural fire regime. And in many water-challenged forests, these small trees can carry fire like a candlewick up to the tops of ancient trees. "Thinning" the small trees can make existing ponderosa pine forests more resilient to wildfires, and to drought and higher temperatures as well. The 2002 Rodeo-Chediski fire that raged through dense White Mountain pine stands, killing trees in city-sized patches, generally laid down and became a much milder surface fire in areas that had been treated by thinning and/or prescribed burns, as Barbara Strom and Peter Fulé of Northern Arizona University documented in 2007. Pine stands treated with thinning and prescribed burns before the Rodeo-Chediski fire swept through also were less likely to convert to oak-manzanita scrub after the fire.

Although the Strom and Fulé study came later, scientific evidence like theirs helped convince a group of northern Arizona locals known as the White Mountains Natural Resources Working Group to push for thinning projects in local forests. The working group of environmentalists, timber entrepreneurs, scientists, agency officials, and politicians collaborated closely with U.S. Forest Service officials to develop policy for this mountainous region of ponderosa pine forests. In 2004, Forest Service managers brought White Mountains Working Group members and environmentalists from the Nature Conservancy, the White Mountain Conservation League, and the Center for Biological Diversity to see firsthand the results of stand-thinning treatments. While the untreated stands contained trees so crowded together they resembled a fine-toothed comb from a distance, the treated stands featured a more open forest that made it easier for people to walk through, yet more difficult for crown fires to navigate. The visitors liked what they saw. In part, their contentment resulted because they had helped decide how to treat these forests. After years of information exchange and negotiation, they had agreed to support thinning projects as long as the cutting left in place any trees with trunks larger than 16 inches in diameter. The visit helped reassure them that the plan was working and the big trees were staying put.

Thinning forests to restore health and resiliency often requires a different approach than logging them to keep sawmills running. Modern sawmills thrive on big trees, whereas it's the "pencil-thin" trees of about a foot in diameter or less that are choking the forest. It actually costs more to harvest the small trees than a typical business can make on them. In fact, the White Mountains Working Group found that it cost about $600 an acre to get these forest stands into shape. But the group decided the goal was worth pursuing. After the devastating 2002 conflagration, the working group rallied other White Mountains residents to support the plan. Treatments can make these pine stands more resistant to the big fires that are projected to increase as the climate warms. Thus, support for such projects can help keep carbon in the woods—by reducing the odds of future forest flambé.

❧

On the other side of the range, the White Mountain Apache Tribe's forest took a similar blow from the 2002 blaze. The Rodeo-Chediski fire scorched about a quarter of a million acres of their reservation land, leaving entire mountainsides bereft of living trees. An estimated 80 million trees died, many of them in city-sized patches. Tribal members soon got to work replanting. By early 2008, ninety-three Apache contractors and workers had planted 1.5 million ponderosa seedlings, reported Mary Stuever, the implementation leader of the tribe's planting efforts. Yet despite the pressing need to get trees in place to reduce soil erosion on these steep slopes, lack of funds forced Stuever to lay off forty-two workers in late 2007. Huge swaths of these mountainous lands remain ominously barren. Meanwhile, the Arizona Department of Commerce's official unemployment rates on the reservation topped 12 percent in 2007, even before the national recession hit. Unofficial rates soar even higher. "The work hasn't gone away. The funding has gone away," Stuever noted during our February 2008 conversation. It's hard to find money beyond the emergency-response years, especially when hurricane-struck cities are among those clamoring for the nation's meager budget for domestic emergencies.

As the funds dwindled, Stuever and others turned their efforts to getting support from groups seeking to offset some of their carbon emissions by planting trees. Super Bowl XLII organizers provided $13,000 for the planting of 42 acres of seedlings in 2008 to help offset the ecological impact of their event at University of Phoenix Stadium that February. When assessing the ecological value of their planting effort, they took into account that probably only a third of the 150 seedlings planted per acre would survive. After all, these are dry times in the Southwest. "Our biggest challenge with tree survival is the same drought conditions that caused the fire," said Stuever, who also describes some of the challenges in her 2009 book, *The Forester's Log*. But if they don't replant, it could take hundreds, even thousands, of years for pines to return to some of these barren lands, she estimated. Other support has been trickling in, but the program continues to struggle. In late 2009, Stuever's replacement, Acting Tribal Forest Manager Jonathan Brooks, told me that he was hoping to secure additional support, including federal stimulus funds, to revamp the planting effort. Replanting trees in fire-damaged White Mountain Apache land seems a classic case of how support for reforestation projects could rescue the landscape from a bleak future and help local people in the process.

Some might question whether it's worthwhile to replant in the face of global warming. Climate models suggest that the southwestern quadrant of the United States overall could face a precipitation decline of 5 to 10 percent along with a temperature rise of 6 to 8 degrees Fahrenheit by the end of the century. Good reasons for replanting remain, however. While the temperature rise could chase pines up the slopes, it's not likely to obliterate pines on Arizona and New Mexico mountains. (Some mountaintop spruce-fir forests, sadly, may be another story.) A 7-degree-Fahrenheit rise would bump ecological niches up by roughly 2,000 feet, based on standard rates of how temperature drops as one scales a mountain. Meanwhile, precipitation remains notoriously difficult to predict at almost any scale, including as it relates to changing climate. If rainfall declines, it certainly could challenge tree survival in some places. Soil moisture plays a key role in seedling survival. But these challenges would be greater in places already deforested than in those still bearing trees, considering the role of forests in retaining and circulating moisture (described in chapter 8).

Keep this in mind, too: If temperature rises as projected, it's because of the extra greenhouse gases in the air. Recall that the main greenhouse gas blamed for the warming, carbon dioxide, also has a well-documented effect of reducing a tree's water needs even while making it more tolerant of higher temperatures (detailed in chapter 5). Also, even if these replanted forest stands were to fail over the centuries, they would still be taking up carbon over the decades while they lasted. What's more, some of the carbon trapped in the wood would resist decay for centuries, and even more of it for decades. In effect, tree-planting projects buy time to give the world's people a chance to wean themselves off fossil fuels. Projects like this also make tribal lands more resilient to climate change while providing jobs for indigenous people. This is important, as the United States' indigenous people are among those most vulnerable to global warming.

❧

Global warming has emerged as a major concern among U.S. indigenous people living on tribal lands. While the boundaries of their reservations remain static, the changing climate is shifting their way of life. "It troubles us because we know the federal government has put boundaries on us. They've nailed us down to the ground," Terry Williams of the Tulalip

Tribe told participants in a 2006 Tribal Lands and Climate conference in Arizona that drew indigenous people representing some fifty tribes ranging from Alaska to Arizona. The Tulalip people have lived in the Puget Sound region of the Pacific Northwest for 10,000 years, he said, but "our culture could be terminated in twenty-five years because the species we need migrate."

Even with these extra challenges, tribes with officially recognized claims to land have it better than those lacking these rights, such as Alaska's Inuit people. In addition to living around the Arctic Circle where the warming is most extreme, they are also more vulnerable to climate change because they must do their hunting on land that is open to the public. This wouldn't be so bad if hunters from out of town followed the same dictates, Hazel Smith, a member of the Inuit tribe, said over lunch at the 2006 conference. "We grew up knowing to let the first of the herd pass," explained Smith. Now federal agencies are requiring that everyone in the country have equal access to hunting on the public lands near the village, such as the Kobuk Valley National Park. "Fine, but don't divert our caribou herds," Smith said. Too often, non-Native hunters will go after the first caribou they see, an unfortunate move that can send the whole herd in another direction. Trying to locate a herd that has strayed off its usual migratory route is no easy feat in the open country of Alaska. With the price of gas to power snowmobiles reaching ten dollars a gallon in some villages, it can get expensive to track down a stray herd—especially for those hunting for subsistence rather than sport. Climate change adds a compounding factor to an already challenging situation. "Our animals don't know any boundaries. They go where the food is," Smith explained. "We survive with the animals."

When considering where to focus on carbon uptake, companies and governments would do well to consider locating projects in areas that could help balance out the past abuse and future uncertainty faced by our indigenous population. For tribes that don't have land, this could provide an ideal opportunity to help right that wrong. For tribes that already have land, projects could be designed that effectively extend the borders of their reservations—preferably in a direction most likely to maintain a climate similar to the one on their existing reservation. Probably this would be to the north, but it would depend on the lay of the land, such as changes in elevation. As with any carbon project, efforts like these would require collaboration with the surrounding community as well as the tribes.

In some cases, improving the ability of tribes to respond to climate change involves merely revising some of the many federal regulations that govern their activities. Federal restrictions applied to tribes limit their ability to respond to changing climate and migrating species, members of several different tribes said. Craig Fleener, a member of the Vuntut Gwitchin Tribe of Alaska's Yukon region, expressed this concern at a 2008 conference in Boulder, Colorado. "We've been adapting to changes forever," Fleener said. "Nowadays, we're set to specific schedules that say 'You have to hunt within these ten days.' The government should allow us to adapt to changes in ways we see would meet our needs the best." Shannon McNeeley, a visiting scientist with the National Center for Atmospheric Research who has been working with Alaskan tribes, presented climate data that supported these concerns. Prime hunting season in the Yukon occurs once daytime temperatures drop into the forties and the nights hit freezing, indigenous people informed her. In recent years, McNeeley said, "this is right when the [hunting] season cuts off. Right when it starts to get good."

Some of the same ideas that would help indigenous people respond to ongoing changes could apply to other groups in vulnerable situations. Hurricane Katrina made it clear that African American communities in Louisiana suffered the most from the storm (with problems worsened by the irresponsible emergency response). In fact, African American communities all over the South face similar risks from future hurricanes, noted Robert Verchick, a law professor at Loyola University in New Orleans and author of *Facing Catastrophe: Environmental Action for a Post-Katrina World.* "Look where African Americans congregate, for a variety of historical reasons going back hundreds of years—they're in the hurricane belt," Verchick said during a 2009 talk at a Tucson conference on adapting to climate change. "Hurricane policy in the United States is by definition African American policy . . . because hurricanes disproportionately will affect African Americans." Meanwhile, people from Hispanic cultures tend to congregate in the U.S. Southwest, where already scorching summer temperatures are slated to rise faster than anywhere else in the country. Across the United States, minority groups disproportionately inhabit barren inner-city neighborhoods most at risk from heat waves—in the Bronx as well as Phoenix. Different areas and ethnic groups have different issues, but many of the Americans most vulnerable to global warming would benefit from local efforts to restore protective wetlands and plant trees where they can thrive, from mountaintops to city plots.

❦

It would take a lot of time and effort to set up a reality-based system that recognizes forests, wetlands, and other natural systems for their climate-control services. It's not a quick fix. Still, done correctly, the protection and restoration of forests and swamps would offer a real answer to the question, What would Gaia do? As we've seen, the planet responded to earlier warming episodes by expanding its forests and wetlands. Not only do these natural systems take up some of the greenhouse gases behind the warming, they also help protect the planet from some of the impacts of climate change. They act as an herbal remedy to the planetary fever we call global warming. Restoring forests to more natural conditions, whether by planting seedlings or thinning out small trees that would have been killed in a natural fire regime, is another valuable service that increases a forest's chance of surviving climate change. It makes sense to find financial support for these worthwhile projects, whether from national budgets targeted for job creation, private funding to support the carbon drawdown, or local dollars in recognition of the many valuable services provided by natural systems in the community.

From a Gaian point of view, restoration efforts should involve species native to the region. Along with having a variety of species, care should be taken to promote genetic diversity (a variety of genotypes) within each species. In fact, a greater diversity of species and genotypes appears to allow for more precise fine-tuning of environmental conditions. Although the importance of species diversity fell mostly between the lines of this book, it could easily comprise another one. Many Gaian thinkers have documented the value of biological diversity to climate control and other functions. One way to ensure that less obvious values such as biological diversity don't get left out of the picture when considering restoration projects is to have a diverse group of people involved in hatching the plans. Groups with active members from environmental nongovernmental organizations will have built-in watchdogs for these types of values. The approach taken by the White Mountains Natural Resources Working Group, described in more detail in my 2006 *Environment* article, offers an example worth imitating. In particular, group members pointed to an open-door policy as crucial to their success.

A national program focused on carbon collection would require oversight by an objective national group with additional oversight from citizens. The U.S. General Accounting Office comes to mind as a neutral trustworthy agency. As of publication, there was no national organization to verify officially whether carbon-offset projects are living up to posted promises and to ensure that credits are sold only once. Often, companies buying offset credits will describe their actual projects, which helps reduce the risk of oversell. Still, a viable system requires a trustworthy oversight group with real authority. In addition, there should be an advisory group with representatives from all interested sectors, including environmental nongovernmental organizations, commercial and industrial businesses, indigenous tribes, and academic researchers as well as government officials.

A variety of forest and wetland restoration projects—on tribal lands, public lands, and private land—could restore key systems and their many life-supporting services. Dialogue and trust building will be important elements of implementing local projects. This will take time. True, time is running short for changing the direction of the current greenhouse-gas trajectory. That's why it's important to start getting the groundwork into place now, even while gearing up for major changes on the energy front. With a good mix of people who have developed trust, carbon projects could green the landscape and provide jobs, even while performing a variety of other valuable services, including moderating local temperatures and moisture, cleaning toxins, and sheltering diverse plants and animals, including humans. The more we can count on forests and wetlands to stabilize the carbon dioxide drawdown, the less pressure we'll put on oceans to take up this greenhouse gas and thus increase their acidity. And the more we pull down greenhouse gases into forests, their soils, and wetlands, the less need the planet will have for the cooling power of hurricanes and floods.

It won't be easy to make the changes needed to address global-warming issues from a Gaian perspective, even with the non-energy focus emphasized here. There's no simple answer to this complex problem. But there is a simplicity to the approach in that the benefits generally apply on many levels. What feels better locally tends to improve the situation globally. Life gets better at all scales when we boost Gaia's natural defenses. Where trees thrive, rivers flow. Where mountains reign, moisture settles. Where marshes sit, water becomes pure. While some could view following the lead of the planet as a way to "Save the Earth," we're really talking about making our own lives better. Won't we all breathe a little easier knowing that our life-support system isn't teetering on the edge of a precipice? We must ask, then: What can we do for our planet and, therefore, ourselves?

A Note on Sources

The documentation sections of this book are divided into Suggested Reading and Works Cited.

Suggested Reading is a list of books, reports, and a lay article that general readers with or without a scientific bent may enjoy. The Works Cited section gives the published sources—print, online, or cinematic—of researchers and experts on the topics mentioned in each chapter.

Scientists, scholars, and other interested readers who would like to know more details can access additional documentation in note form, presented by page number and last phrase of the sentence, at the following Web site: http://www.uapress.arizona.edu/extras/Lenart/

Suggested Reading

Aguado, E., and J. E. Burt, 1999. *Understanding Weather and Climate*. Upper Saddle River, NJ: Prentice Hall.

Bader, D. C., C. Covey, W. J. Gutowski, I. M. Held, K. E. Kunkel, R. L. Miller, R. T. Tokmakian, and M. H. Zhang, 2008. *Climate Models: An Assessment of Strengths and Limitations*. A report by the U.S. Climate Change Science Program and the Subcommittee on Global Change Research. Washington, D.C.: Department of Energy, Office of Biological and Environmental Research.

Bowen, M., 2005. *Thin Ice: Unlocking the Secrets of Climate in the World's Highest Mountain*. New York: Henry Holt.

Brown, L. R., 2008. *Plan* $B_{3.0}$: *Mobilizing to Save Civilization*. New York: W. W. Norton.

Burt, C. C., 2004. *Extreme Weather: A Guide and Record Book*. New York: W. W. Norton.

Cajete, G., 2000. *Native Science: Natural Laws of Interdependence*. Santa Fe, NM: Clear Light Publishers.

Emanuel, K., 2005. *Divine Wind: The History and Science of Hurricanes*. New York: Oxford University Press.

Flannery, T., 2005. *The Weather Makers: How Man Is Changing the Climate and What It Means for Life on Earth*. New York: Atlantic Monthly Press.

Glantz, M. H., 2003. *Climate Affairs: A Primer*. Washington, D.C.: Island Press.

Gore, A., 2006. *An Inconvenient Truth: The Planetary Emergency of Global Warming and What We Can Do about It*. New York: Rodale and Melcher Media.

Greb, S. F., and W. A. DiMichele (eds.), 2006. *Wetlands through Time*. Boulder, CO: Geological Society of America Special Paper 399.

Harding, S., 2006. *Animate Earth: Science, Intuition and Gaia*. White River Junction, VT: Chelsea Green Publishing Company.

Hill, J. B., 2000. *The Legacy of Luna: The Story of a Tree, a Woman, and the Struggle to Save the Redwoods*, pages 76–77. San Francisco: HarperSanFrancisco (division of HarperCollins).

Intergovernmental Panel on Climate Change (IPCC). Reports are available online at www.ipcc.ch.

Kenrich, P., and P. Davis, 2004. *Fossil Plants*. Washington, D.C.: Smithsonian Books in association with the Natural History Museum, London.

Lenart, M., with contributions from G. Garfin, B. Colby, T. Swetnam, B. J. Morehouse, S. Doster, and H. Hartmann, 2007. *Global Warming in the Southwest: Projections, Observations and Impacts*. Tucson, AZ: Climate Assessment for the Southwest (University of Arizona). Online at http://www.climas.arizona.edu/pubs/pdfs/GWSouthwest.pdf.

Lovelock, J. E., 1979. *Gaia: A New Look at Life on Earth*. Oxford: Oxford University Press.

Margulis, L., 1998. *Symbiotic Planet: A New Look at Evolution*. New York: Basic Books.

Maser, C., and J. R. Sedell, 1994. *From the Forest to the Sea: The Ecology of Wood in Streams, Rivers, Estuaries, and Oceans*. Delray Beach, FL: St. Lucie Press.

McKibben, B., 2006. *The End of Nature*. New York: Random House.

McKibben, B., 2007. *Fight Global Warming Now: The Handbook for Taking Action in Your Community*. New York: Henry Holt. For more on the 350 campaign, see http://www.350.org.

Mitsch, W. J., and J. G. Gosselink, 2007. *Wetlands* (4th ed.). Hoboken, NJ: John Wiley & Sons.

Mooney, C., 2007. *Storm World: Hurricanes, Politics and the Battle over Global Warming*. Orlando, FL: Harcourt, Inc.

Murname, R. J., and Kam-Biu Liu, 2004. *Hurricanes and Typhoons: Past, Present, and Future*. New York: Columbia University Press.

Ruddiman, W. F., 2001. *Earth's Climate, Past and Future*. New York: W. H. Freeman.

Schneider, S. H., J. R. Miller, E. Crist, and P. J. Boston. *Scientists Debate Gaia: The Next Century*. Cambridge, MA: MIT Press.

Shah, S., 2004. *Crude: The Story of Oil*. New York: Seven Stories Press.

Skelton, P. W., Spicer, R. A., S. P. Kelley, and I. Gilmour, 2003. *The Cretaceous World*. Cambridge, U.K.: Cambridge University Press.

Stuever, M., 2009. *The Forester's Log: Musings from the Woods*. Albuquerque: University of New Mexico Press.

Tidwell, M., 2006. *The Ravaging Tide: Strange Weather, Future Katrinas, and the Coming Death of America's Coastal Cities*. New York: Free Press (division of Simon & Schuster).

Trenberth, K. E., 2007. Warmer oceans, stronger hurricanes. *Scientific American* July:44–51.

Ward, P. D., 2006. *Out of Thin Air: Dinosaurs, Birds and Earth's Ancient Atmosphere*. Washington, D.C.: Joseph Henry Press.

Weisman, A., 2007. *The World without Us*. New York: Thomas Dunne Books (imprint of St. Martin's Press).

Works Cited

Introduction

Abrams, D., 1991. The mechanical and the organic: On the impact of metaphor in science. In *Scientists on Gaia* (Stephen Schneider and Penelope Boston, eds.), pages 66–77. Cambridge, MA: MIT Press.

Brazel, A., N. Selover, R. Vose, and G. Heisler, 2000. The tale of two climates—Baltimore and Phoenix urban LTER sites, *Climate Research* 15:123–135 (July 20).

Intergovernmental Panel on Climate Change (IPCC), 2007. *Climate Change 2007: The Physical Science Basis. Contribution of Working Group I to the Fourth Assessment Report of the Intergovernmental Panel on Climate Change* (S. Solomon, D. Qin, M. Manning, Z. Chen, M. Marquis, K. B. Averyt, M. Tignor, and H. L. Miller, eds.). Cambridge, U.K., and New York: Cambridge University Press.

Lovelock, J. E., 2006. *The Revenge of Gaia: Why the Earth Is Fighting Back—And How We Can Still Save Humanity*. London: Allen Lane (imprint of Penguin Books).

Margulis, L., 1998. *Symbiotic Planet: A New Look at Evolution*, pages 127–128. New York: Basic Books.

Petit, J.-R., J. Jouzel, D. Raynaud, N. I. Barkov, J. M. Barnola, I. Basile, M. Bender, J. Chappellaz, M. Davis, G. Delaygue, M. Delmotte, V. M. Kotlyakov, M. Legrand, V. Y. Lipenkov, C. Lorius, L. Pepin, C. Ritz, E. Saltzman, and M. Stievenard, 1999. Climate and atmospheric history of the past 420,000 years from the Vostok ice core, Antarctica. *Nature* 399:429–436.

Westerling, A. L., H. G. Hidalgo, D. R. Cayan, and T. W. Swetnam, 2006. Warming and earlier spring increase western U.S. forest wildfire activity. *Science* 313:940–943.

Chapter 1. A Feverish Response

Bender, M. A., and I. Ginis, 2000. Real-case simulations of hurricane-ocean interaction using a high-resolution coupled model: Effects on hurricane intensity. *Monthly Weather Review* 128:917–945.

Bender, M. A., I. Ginis, and Y. Kurihara, 1993. Numerical simulations of tropical cyclone-ocean interaction with a high-resolution coupled model. *Journal of Geophysical Research* 98:23,245–23,263.

Bice, K. A., D. Birgel, P. A. Meyers, K. A. Dahl, K. U. Hinrichs, and R. D. Norris, 2006. A multiple proxy and model study of Cretaceous upper ocean temperatures and atmospheric CO_2 concentrations. *Paleoceanography* 21(PA2002):1–17.

Cerveny, R. S., and L. E. Newman, 2000. Climatological relationships between tropical cyclones and rainfall. *Monthly Weather Review* 128:3329–3336.

Davis, A., and X.-H. Yan, 2004. Hurricane forcing on chlorophyll-a concentration off the northeast coast of the U.S. *Geophysical Research Letters* 31:L17304.

Emanuel, K., 1999. Thermodynamic control of hurricane intensity. *Nature* 410:665–669.

Emanuel, K., 2001. Contribution of tropical cyclones to meridional heat transport by the oceans. *Journal of Geophysical Research* 106(D14):14,771–14,781.

Emanuel, K., 2005. Increasing destructiveness of tropical cyclones over the past 30 years. *Nature* 436:686–688.

Emanuel, K., R. Sundararajan, and J. Williams, 2008. Hurricanes and global warming: Results from downscaling IPCC AR4 simulations. *Bulletin of the American Meteorological Society* 89(3):347–367.

Goldenberg, S. B., C. W. Landsea, A. M. Mestas-Nuñez, and W. M. Gray, 2001. The recent increase in Atlantic hurricane activity: Causes and implications. *Science* 293(20 July):474–478.

Gray, W. M., 1968. A global view of the origin of tropical disturbances and storms. *Monthly Weather Review* 96:669–700.

Hobgood, J. S., and R. S. Cerveny, 1988. Ice-age hurricanes and tropical storms. *Nature* 333:243–245.

Ito, M., A. Ishigaki, T. Nishikawa, and T. Saito, 2001. Temporal variation in the wavelength of hummocky cross-stratification: Implications for storm intensity through Mesozoic and Cenozoic. *Geology* 29:87–89.

Kang, W.-J., and J. H. Trefry, 2003. Retrospective analysis of the impacts of major hurricanes on sediments in the lower Everglades and Florida Bay. *Environmental Geology* 44:771–780.

Knutson, T. R., and R. E. Tuleya, 2004. Impact of CO_2-induced warming on simulated hurricane intensity and precipitation: Sensitivity to the choice of climate model and convective parameterization. *Journal of Climate* 17(18):3477–3495.

Landsea, C. W., G. A. Vecchi, L. Bengtsson, and T. R. Knutson (in press). Impact of duration thresholds on Atlantic tropical cyclone counts. *Journal of Climate*. Submission online at ftp://ftp.gfdl.gov/pub/gav/PAPERS/LVBK_08_SHORTSTORMS.submitted.pdf.

Liu, K., 2004. Paleotempestology: Principles, methods, and examples from Gulf Coast sediments. In *Typhoons: Past, Present, and Future* (R. Murname and K. Liu, eds.), pages 13–57. New York: Columbia University Press.

Mann, M. E., and K. E. Emanuel, 2006. Atlantic hurricane trends linked to climate change. *Eos: Transactions of the American Geophysical Union* 87(24):233–244.

Mann, M. E., J. D. Woodruff, J. P. Donnelly, and Z. Zhang, 2009. Atlantic hurricanes and climate over the past 1,500 years. *Nature* 460:880–883.

Michaels, P. J., P. C. Knappenberger, and R. E. Davis, 2006. Sea-surface temperatures and tropical cyclones in the Atlantic basin. *Geophysical Research Letters* 33(L09708):1–4.

Mooney, C., 2007. *Storm World: Hurricanes, Politics and the Battle over Global Warming*. Orlando, FL: Harcourt, Inc.

National Oceanic and Atmospheric Administration, August 11, 2009, press release. Study: Better observations, analyses detecting short-lived tropical systems. Posted at http://www.noaanews.noaa.gov/stories2009/20090811_tropical.html.

Pearson, P. N., P. W. Ditchfield, J. Singano, K. G. Harcourt-Brown, C. J. Nicholas, R. K. Olsson, N. J. Shackleton, and M. A. Hall, 2001. Warm tropical sea surface temperatures in the late Cretaceous and Eocene epochs. *Nature* 413:481–487.

Pearson, P. N., B. E. van Dongen, C. J. Nicholas, R. D. Pancost, S. Schouten, J. M. Singano, and B. W. Wade, 2007. Stable warm tropical climate through the Eocene epoch. *Geology* 35(3):211–214.

Sanford, R. L., Jr., W. J. Parton, D. S. Ojima, and D. J. Lodge, 1991. Hurricane effects on soil organic matter dynamics and forest production in the Luquillo Experimental Forest, Puerto Rico: Results of simulation modeling. *Biotropica* 23(4a):364–372.

Scatena, F. N., and M. C. Larsen, 1991. Physical aspects of Hurricane Hugo in Puerto Rico. *Biotropica* 23(4a):317–323.

Scatena, F. N., S. Moya, C. Estrada, and J. D. Chinea, 1996. The first five years in the reorganization of aboveground biomass and nutrient use following Hurricane Hugo in the Bisley Experimental Watersheds, Luquillo Experimental Forest, Puerto Rico. *Biotropica* 28(4a):424–440.

Shay, L. K., G. J. Goni, and P. G. Black, 2000. Effects of a warm oceanic feature on Hurricane Opal. *Monthly Weather Review* 128:1366–1383.

Sriver, R., and M. Huber, 2006. Low-frequency variability in globally integrated tropical cyclone power dissipation. *Geophysical Research Letters* 33:L11705.

Sriver, R., and M. Huber, 2007. Observational evidence for an ocean heat pump induced by tropical cyclones. *Nature* 447:577–580.

Sriver, R. L., M. Huber, and J. Nusbaumer, 2008. Investigating tropical cyclone-climate feedbacks using the TRMM Microwave Imager and Quick Scatterometer. *Geochemistry, Geophysics, Geosystems* 9:Q09V11.

Trenberth, K. E., 2007. Warmer oceans, stronger hurricanes. *Scientific American* July:44–51.

Trenberth, K. E., and J. Fasullo, 2008. Energy budgets of Atlantic hurricanes and changes from 1970. *Geochemistry, Geophysics, Geosystems* 9:Q09V08, doi:10.1029/2007GC001847.

Vecchi, G. A., and B. J. Soden, 2007. Increased tropical Atlantic wind shear in model projections of global warming. *Geophysical Research Letters* 34:L08702.

Walker, L. R., 1991. Tree damage and recovery from Hurricane Hugo in Luquillo Experimental Forest, Puerto Rico. *Biotropica* 23(4A):379–385.

Wang, C., S. K. Lee, and D. B. Enfield, 2008. Atlantic warm pool acting as a link between Atlantic multidecadal oscillation and Atlantic tropical cyclone activity. *Geochemistry, Geophysics, Geosystems* 9(5):1–17.

Webster, P. J., G. J. Holland, J. A. Curry, and H. R. Chang, 2005. Changes in tropical cyclone number, duration and intensity in a warming environment. *Science* 309(5742):1844–1846.

Yih, K., D. H. Boucher, J. H. Vandermeer, and N. Zamora, 1991. Recovery of the rain forest of southeastern Nicaragua after destruction by Hurricane Joan. *Biotropica* 23(2):106–113.

Chapter 2. A Living System

Abrams, D., 1991. The mechanical and the organic: On the impact of metaphor in science. In *Scientists on Gaia* (Stephen Schneider and Penelope Boston, eds.), pages 66–77. Cambridge, MA: MIT Press.

Cajete, G., 2000. *Native Science: Natural Laws of Interdependence*. Santa Fe, NM: Clear Light Publishers.

Chopra, Deepak, 1993. *Ageless Body, Timeless Mind: The Quantum Alternative to Growing Old*. New York: Harmony Books.

Deloria, V., Jr., 1969. *Custer Died for Your Sins: An Indian Manifesto*. New York: Macmillan.

Deloria, V., Jr., and D. R. Wildcat, 2001. *Power and Place: Indian Education in America*. Golden, CO: Fulcrum Resources.

Emanuel, K., 2002. A simple model of multiple climate regimes. *Journal of Geophysical Research–Atmospheres* 107(D9):4077, 1–10.

Kirchner, J. W., 1991. The Gaia hypotheses: Are they testable? Are they useful? In *Scientists on Gaia* (S. H. Schneider and P. J. Boston, eds.), pages 38–46. Cambridge, MA: MIT Press.

Lovelock, J. E., 1979. *Gaia: A New Look at Life on Earth*, page 11. Oxford: Oxford University Press.

Lovelock, J. E., 2006. *The Revenge of Gaia: Why the Earth Is Fighting Back—And How We Can Still Save Humanity*. London: Allen Lane (imprint of Penguin Books).

Lovelock, J. E., and L. Margulis, 1974. Atmospheric homeostasis by and for the biosphere: The gaia hypothesis. *Tellus* 24(1–2):2–9.

Lovelock, J. E., and A. J. Watson, 1982. The regulation of carbon dioxide and climate: Gaia or geochemistry. *Planetary and Space Science* 30:795–802.

Margulis, L., 1993. *Symbiosis in Cell Evolution: Microbial Communities in the Archean and Proterozoic Eons* (2nd ed.). New York: W. H. Freeman.

Margulis, L., 1998. *Symbiotic Planet: A New Look at Evolution*, pages 127–128. New York: Basic Books.

Margulis, L., and M. J. Chapman, 2009. *Kingdoms and Domains: Illustrated Phyla of Life in Color*. Amsterdam: Elsevier.

Margulis, L., and J. E. Lovelock, 1974. Biological modulation of the Earth's atmosphere. *Icarus* 21:471–489.

Odum, E. P., in collaboration with H. T. Odum, 1959. *Fundamentals of Ecology* (2nd ed.), pages vi and 16. Philadelphia: W. B. Saunders.

Sagan, C., and G. Mullen, 1972. Earth and Mars: Evolution of atmospheres and surface temperatures. *Science* 177 (4043):52–56.

Scofield, Bruce, 2004. Gaia: The Living Earth—2,500 years of precedents in natural science and philosophy. In *Scientists Debate Gaia: The Next Century* (S. H. Schneider, J. R. Miller, E. Crist, and P. J. Boston, eds.), pages 151–160. Cambridge, MA: MIT Press.

Thomas, L., 1974. *The Lives of a Cell: Notes of a Biology Watcher*, page 145. New York: Viking Press.

Turner, J. S., 2006. *The Tinkerer's Accomplice: How Design Emerges from Life Itself*. Cambridge, MA: Harvard University Press.

Chapter 3. Greenhouse-Gas Attack

Andreae, M. O., 2007. Atmospheric aerosols versus greenhouse gases in the twenty-first century. *Philosophical Transactions of the Royal Society A* 365:1915–1923.

Andreae, M. O., C. D. Jones, and P. M. Cox, 2005. Strong present-day aerosol cooling implies a hot future. *Nature* 435:1187–1190.

Berner, R. A., 2004. *The Phanerozoic Carbon Cycle: CO_2 and O_2*. Oxford: Oxford University Press.

Blunier, T., E. Monnin, and J. M. Barnola, 2005. Atmospheric data from ice cores: Four climatic cycles. In *A History of Atmospheric CO_2 and Its Effects on Plants, Animals and Ecosystems* (J. R. Ehleringer, M. D. Dearing, and T. E. Cerling, eds.), pages 62–82. Berlin: Springer.

Budyko, M. I., 1982. *The Earth's Climate: Past and Future*. New York: Academic Press, International Geophysics Series.

Cerling, T. E., 1989. Does the gas content of amber reveal the composition of palaeoatmospheres? *Nature* 339:695–696.

Chappellaz, J., D. Luethi, L. Loulergue, J. Barnola, B. Bereiter, T. Blunier, J. Jouzel, M. Lefloch, B. Lemieux, V. Masson-Delmotte, D. Raynaud, A. France Schilt, U. Siegenthaler, R. Spahni, and T. Stocker, 2007. Greenhouse gas concentration records extended back to 800,000 years from the EPICA Dome C Ice Core. *Eos Transaction*, AGU 88(52), Fall Meeting Supplement, Abstract PP31E-01.

DeCicco, J., F. Fung, and F. An, 2006. Global Warming on the Road: The Climate

Impact of America's Automobiles. Washington, D.C.: Environmental Defense Fund. Online at http://www.edf.org/documents/5301_Globalwarmingontheroad.pdf.

DeConto, R. M., and D. Pollard, 2003. Rapid Cenozoic glaciation of Antarctica induced by declining atmospheric CO_2. *Nature* 421:245–249.

Emanuel, K., 2001. Contribution of tropical cyclones to meridional heat transport by the oceans. *Journal of Geophysical Research* 106(D14):14,771–14,781.

Eugster, W., W. Rouse, R. Pielke, J. P. McFadden, D. D. Baldocchi, T. Kittel, F. S. Chapin III, G. E. Liston, P. L. Vidale, E. A. Vaganov, and S. Chambers, 2000. Land-atmosphere energy exchange in Arctic tundra and boreal forest: Available data and feedbacks to climate. *Global Change Biology* 6:84–115.

Guggenheim, Davis (director), 2006. *An Inconvenient Truth: A Global Warning*. Global Warming Documentary, LLC, and Paramount Pictures.

Hansen, J., 2007. Huge sea level rises are coming—unless we act now. *New Scientist Environment*, initially published July 25, 2007, by NewScientist.com news service.

Hays, J., J. Imbrie, and N. Shackleton, 1976. Variations in the Earth's orbit: Pacemaker of the ice ages. *Science* 194: 1121–1132.

Huber, M., and L. Sloan, 2000. Climatic responses to tropical sea surface temperature changes on a "greenhouse" Earth. *Paleoceanography* 15:443–450.

Imbrie, J., and K. Palmer Imbrie, 1979. *Ice Ages: Solving the Mystery*. Short Hills, NJ: Enslow Publishers.

Intergovernmental Panel on Climate Change (IPCC), 1995. *Climate Change 1995: The Science of Climate Change* (J. T. Houghton, L. G. Meiro Filho, B. A. Callender, N. Harris, A. Kattenburg, and K. Maskell, eds.). Cambridge, U.K., and New York: Cambridge University Press.

Intergovernmental Panel on Climate Change (IPCC), 2001. *Climate Change 2001: The Scientific Basis. Contribution of Working Group I to the Third Assessment Report of the Intergovernmental Panel on Climate Change* (J. T. Houghton, Y. Ding, D. J. Griggs, M. Noguer, P. J. van der Linden, X. Dai, K. Maskell, and C. A. Johnson, eds.). Cambridge, U.K., and New York: Cambridge University Press.

Intergovernmental Panel on Climate Change (IPCC), 2007. *Climate Change 2007: The Physical Science Basis. Contribution of Working Group I to the Fourth Assessment Report of the Intergovernmental Panel on Climate Change* (S. Solomon, D. Qin, M. Manning, Z. Chen, M. Marquis, K. B. Averyt, M. Tignor, and H. L. Miller, eds.). Cambridge, U.K., and New York: Cambridge University Press.

Jahren, A. H., N. C. Arens, G. Sarmiento, J. Guerrero, and R. Amundson, 2001. Terrestrial record of methane hydrate dissociation in the early Cretaceous. *Geology* 29(2):159–162.

Jouzel, J., V. Masson-Delmotte, O. Cattani, G. Dreyfus, S. Falourd, G. Hoffmann, B. Minster, J. Nouet, J. M. Barnola, J. Chappellaz, H. Fischer, J. C. Gallet,

S. Johnsen, M. Leuenberger, L. Loulergue, D. Luethi, H. Oerter, F. Parrenin, G. Raisbeck, D. Raynaud, A. Schilt, J. Schwander, E. Selmo, R. Souchez, R. Spahni, B. Stauffer, J. P. Steffensen, B. Stenni, T. F. Stocker, J. L. Tison, M. Werner, and E. W. Wolff, 2007. Orbital and millennial Antarctic climate variability over the past 800,000 years. *Science* 317:793–796.

Kaufman, D. S., D. P. Schneider, N. P. McKay, C. M. Ammann, R. S. Bradley, K. R. Briffa, G. H. Miller, B. L. Otto-Bliesner, J. T. Overpeck, B. M. Vinther, and Arctic Lakes 2k Project Members, 2009. Recent warming reverses long-term Arctic cooling. *Science* 325:1236–1239.

Keeling, C. D., S. C. Piper, R. B. Bacastow, M. Wahlen, T. P. Whorf, M. Heimann, and H. A. Meijer, 2001. *Exchanges of Atmospheric* CO_2 *and* $^{13}CO_2$ *with the Terrestrial Biosphere and Oceans from 1978 to 2000. I: Global Aspects.* San Diego: Scripps Institution of Oceanography, SIO Reference Series, No. 01-06.

Kukla, G. J., M. L. Bender, J. L. de Beaulieu, G. Bond, W. S. Broecker, P. Cleveringa, J. E. Gavin, T. D. Herbert, J. Imbrie, J. Jouzel, L. D. Keigwin, K. L. Knudsen, J. F. McManus, J. Merkt, D. R. Muhs, and H. Müller, 2002. Last interglacial climates. *Quaternary Research* 58:2–13.

Mann, M. E., R. S. Bradley, and M. K. Hughes, 1999. Northern Hemisphere temperatures during the past millennium: Inferences, uncertainties and limitations. *Geophysical Research Letters* 26(6):759–762.

Mann, M. E., M. A. Cane, S. E. Zebiak, and A. Clement, 2005. Volcanic and solar forcing of the tropical Pacific over the past 1000 years. *Journal of Climate* 18:447–456.

Markwick, P. J., 1994. "Equability," continentality, and Tertiary "climate": The crocodilian perspective. *Geology* 22:613–616.

Milankovitch, M. M., 1941. Canon of insolation and the ice-age problem. Beograd: Koniglich Serbische Akademie. English translation by the Israel Program for Scientific Translations, published for the U.S. Department of Commerce and the National Science Foundation, Washington, D.C., 1969.

Miller, K. G., M. A. Kominz, J. V. Browning, J. D. Wright, G. S. Mountain, M. E. Katz, P. J. Sugarman, B. S. Cramer, N. Christie-Blick, and S. F. Pekar, 2005. The Phanerozoic record of global sea level change. *Science* 310:1293–1298.

Mishchenko, M. I., I. V. Geogdzhayev, W. B. Rossow, B. Cairns, B. E. Carlson, A. A. Lacis, L. Liu, and L. D. Travis, 2007. Long-term satellite record reveals likely recent aerosol trend. *Science* 315:1543.

NOVA, 2006. *Dimming the Sun.* Online at http://www.pbs.org/wgbh/nova/sun.

Overpeck, J. T., B. L. Otto-Bliesner, G. H. Miller, D. R. Muhs, R. B. Alley, and J. T. Kiehl, 2006. Paleoclimatic evidence for future ice-sheet instability and rapid sea-level rise. *Science* 311:1747–1750.

Pearson, P. N., P. W. Ditchfield, J. Singano, K. G. Harcourt-Brown, C. J. Nicholas, R. K. Olsson, N. J. Shackleton, and M. A. Hall, 2001. Warm tropical sea surface temperatures in the late Cretaceous and Eocene epochs. *Nature* 413:481–487.

Pearson, P. N., B. E. van Dongen, C. J. Nicholas, R. D. Pancost, S. Schouten, J. M. Singano, and B. W. Wade, 2007. Stable warm tropical climate through the Eocene epoch. *Geology* 35(3):211–214.

Petit, J.-R., J. Jouzel, D. Raynaud, N. I. Barkov, J. M. Barnola, I. Basile, M. Bender, J. Chappellaz, M. Davis, G. Delaygue, M. Delmotte, V. M. Kotlyakov, M. Legrand, V. Y. Lipenkov, C. Lorius, L. Pepin, C. Ritz, E. Saltzman, and M. Stievenard, 1999. Climate and atmospheric history of the past 420,000 years from the Vostok ice core, Antarctica. *Nature* 399:429–436.

Royer, D. L., 2006. CO_2-forced climate thresholds during the Phanerozoic. *Geochimica et Cosmochimica Acta* 70:5665–5675.

Royer, D. L., R. A. Berner, and J. Park, 2007. Climate sensitivity constrained by CO_2 concentrations over the past 420 million years. *Nature* 446:530–532.

Shackleton, N. J., 2000. The 100,000-year ice-age cycle identified and found to lag temperature, carbon dioxide and orbital eccentricity. *Science* 289:1897–1902.

Skelton, P. W., R. A. Spicer, S. P. Kelley, and I. Gilmour, 2003. *The Cretaceous World* (P. W. Skelton, ed.). Cambridge, U.K.: The Open University and Cambridge University Press.

Smith, T. M., and R. W. Reynolds, 2005. A global merged land, air, and sea surface temperature reconstruction based on historical observations (1880–1997). *Journal of Climate* 18:2021–2036.

Thomas, E., J. C. Zachos, and T. J. Bralower, 2000. Deep-sea environments on a warm Earth: Latest Paleocene–early Eocene. In *Warm Climates in Earth History* (B. T. Huber, K. G. MacLeod, and S. L. Wing, eds.), pages 132–160. Cambridge, U.K.: Cambridge University Press.

Thompson, L. G., E. Mosley-Thompson, M. E. Davis, K. Henderson, H. H. Brecher, V. S. Zagorodnov, T. A. Mashiotta, P. N. Lin, V. N. Mikhalenko, D. R. Hardy, and J. Beer, 2002. Kilimanjaro ice core records: Evidence of Holocene climate change in tropical Africa. *Science* 298:589–593.

Zachos, J., M. Pagani, L. Sloan, E. Thomas, and K. Billups, 2001. Trends, rhythms, and aberrations in global climate 65 Ma to present. *Science* 292:686–693.

Chapter 4. Circulation Patterns

Blanford, H. F., 1884. On the connexion of the Himalaya snowfall with dry winds and seasons of drought in India. *Proceedings of the Royal Society of London* 37:3–22.

Bowen, M., 2005. *Thin Ice: Unlocking the Secrets of Climate in the World's Highest Mountain.* New York: Henry Holt.

Broecker, W. S., 2002. *The Glacial World according to Wally*. Palisades, NY: Eldigio Press, Lamont-Doherty Earth Observatory of Columbia University.

Burnett, A. W., M. E. Kirby, H. T. Mullins, and W. P. Patterson, 2003. Increasing Great Lake-effect snowfall during the twentieth century: A regional response to global warming? *Journal of Climate* 16:3535–3541.

Goswami, B. N., J. Shukla, E. K. Schneider, and Y. C. Sud, 1984. Study of the dynamics of the Intertropical Convergence Zone with a symmetric version of the GLAS Climate Model. *Journal of the Atmospheric Sciences* 41(1):5–19.

Groisman, P. Y., R. W. Knight, T. R. Karl, D. R. Easterling, B. Sun, and J. H. Lawrimore, 2004. Contemporary changes of the hydrological cycle over the contiguous United States: Trends derived from *in-situ* observations. *Journal of Hydrometeorology* 5:64–85.

Gupta, A. K., D. M. Anderson, and J. T. Overpeck, 2003. Abrupt changes in the Asian southwest monsoon during the Holocene and their links to the North Atlantic Ocean. *Nature* 421:354–357.

Hays, J., J. Imbrie, and N. Shackleton, 1976. Variations in the Earth's orbit: Pacemaker of the ice ages. *Science* 194:1121–1132.

Higgins, R. W., Y. Yao, and X. L. Wang, 1997. Influence of the North American monsoon system on the U.S. summer precipitation regime. *Journal of Climate* 10:2600–2622.

Hu, Y., and Q. Fu, 2007. Observed poleward expansion of the Hadley circulation since 1979. *Atmospheric Chemistry and Physics* 7:5229–5236.

Jahren, A. H., and L. S. L. Sternberg, 2002. Eocene meridional weather patterns reflected in the oxygen isotopes of arctic fossil wood. *GSA Today* 12(1):4–9.

Karl, T., and K. Trenberth, 2003. Modern global climate change. *Science* 302:1719–1723.

Knowles, N., M. D. Dettinger, and D. R. Cayan, 2006. Trends in snowfall versus rainfall in the western United States. *Journal of Climate* 19:4545–4559.

Lancaster, B., 2006. *Rainwater Harvesting for Drylands and Beyond.* Vol. 1: *Guiding Principles to Welcome Rain into Your Life and Landscape*. White River Junction, VT: Rainsource Press, distributed by Chelsea Green Publishing Company.

Mitchell, D., D. Ivanova, R. Rabin, T. Brown, and K. Redmond, 2002. Gulf of California sea surface temperatures and the North American monsoon: Mechanistic implications from observations. *Journal of Climate* 15:2261–2281.

Mote, P. W., A. F. Hamlet, M. P. Clark, and D. P. Lettenmaier, 2005. Declining mountain snowpack in western North America. *Bulletin of the American Meteorological Society* 86:39–49.

Pielke, R. A., Jr., and M. W. Downton, 2000. Precipitation and damaging floods: Trends in the United States, 1932–97. *Journal of Climate* 13:3625–3637.

Prentice, C., P. J. Bartlein, and T. Webb III, 1991. Vegetation and climate change in eastern North America since the Last Glacial Maximum. *Ecology* 72(6):2038–2056.

Probst, J.-L., and Y. Tardy, 1987. Long-range streamflow and world continental runoff fluctuations since the beginning of this century. *Journal of Hydrology* 94:289–311.

Schultz, H., U. von Rad, and H. Erlenkeuser, 1998. Correlation between Arabian Sea and Greenland climate oscillations of the past 110,000 years. *Nature* 393:54–57.

Seidel, D. J., and W. J. Randel, 2007. Recent widening of the tropical belt:

Evidence from tropopause observations. *Journal of Geophysical Research* 112(D20113):1–6.

Seidel, D. J., Q. Fu, W. J. Randel, and T. J. Reichler, 2008. Widening of the tropical belt in a changing climate. *Nature Geoscience* 1:21–24.

Skelton, P. W., R. A. Spicer, S. P. Kelley, and I. Gilmour, 2003. *The Cretaceous World* (P. W. Skelton, ed.). Cambridge, U.K.: The Open University and Cambridge University Press.

Stewart, I. T., D. R. Cayan, and M. D. Dettinger, 2005. Changes toward earlier streamflow timing across Western North America. *Journal of Climate* 18:1136–1155.

Thomas, M. F., and M. B. Thorp, 1996. The response of geomorphic systems to climatic and hydrologic change during the late Glacial and early Holocene in the humid and sub-humid tropics. In *Global Continental Changes: The Context of Palaeohydrology* (J. Branson, A. G. Brown, and K. J. Gregory, eds.), pages 139–153. London: Geological Society Special Publication No. 115.

Trenberth, K. E., 1999. Atmospheric moisture recycling: Role of advection and local evaporation. *Journal of Climate* 12:1368–1381.

Trenberth, K. E., P. D. Jones, P. Ambenje, R. Bojariu, D. Easterling, A. Klein Tank, D. Parker, F. Rahimzadeh, J. A. Renwick, M. Rusticucci, B. Soden, and P. Zhai, 2007. Observations: Surface and atmospheric climate change. In Intergovernmental Panel on Climate Change (IPCC), *Climate Change 2007: The Physical Science Basis. Contribution of Working Group I to the Fourth Assessment Report of the Intergovernmental Panel on Climate Change* (S. Solomon, D. Qin, M. Manning, Z. Chen, M. Marquis, K. B. Averyt, M. Tignor, and H. L. Miller, eds.). Cambridge, U.K., and New York: Cambridge University Press.

Zachos, J., M. Pagani, L. Sloan, E. Thomas, and K. Billups, 2001. Trends, rhythms, and aberrations in global climate 65 Ma to present. *Science* 292:686–693.

Zhang, C., 1993. Large-scale variability of atmospheric deep convection in relation to sea surface temperature in the tropics. *Journal of Climate* 6:1898–1913.

Zhang, X., F. W. Zwiers, G. C. Hegerl, F. H. Lambert, N. P. Gillett, S. Solomon, P. A. Stott, and T. Nozawa, 2007. Detection of human influence on twentieth-century precipitation trends. *Nature* 448:461–465.

Chapter 5. An Herbal Remedy

Beerling, D. J., F. I. Woodward, and P. J. Valdes, 1999. Global terrestrial productivity in the mid-Cretaceous (100 Ma): Model simulations and data. In *Evolution of the Cretaceous Ocean-Climate System* (E. Barrera and C. C. Johnson, eds.). Boulder, CO: Geological Society of America Special Paper 332.

Betts, R. A., P. M. Cox, M. Collins, P. P. Harris, C. Huntingford, and C. D. Jones, 2004. The role of ecosystem-atmosphere interactions in simulated Amazonia precipitation decrease and forest dieback under global climate warming. *Theoretical and Applied Climatology* 78:157–175.

Bice, K. A., D. Birgel, P. A. Meyers, K. A. Dahl, K. U. Hinrichs, and R. D. Norris, 2006. A multiple proxy and model study of Cretaceous upper ocean temperatures and atmospheric CO_2 concentrations. *Paleoceanography* 21(PA2002):1–17.

Cerling, T. E., J. M. Harris, B. J. MacFadden, M. G. Leakey, J. Quade, V. Eisenmann, and J. R. Ehleringer, 1997. Global vegetation change through the Miocene/Pliocene boundary. *Nature* 389:153–158.

Chiras, D. D., 1988. *Environmental Science* (2d ed.). Menlo Park, CA: Benjamin/Cummings Publishing.

Clark, D. A., 2007. Detecting tropical forests' responses to global climatic and atmospheric change: Current challenges and a way forward. *Biotropica* 39(1):4–19.

Coope, G. R., 1974. Interglacial Coleoptera from Bobbitshole, Ipswich. *Journal of the Geological Society of London* 130:333–340.

Coope, G. R., 1977. Quaternary Coleoptera as aids in the interpretation of environmental history. In *British Quaternary Studies* (F. W. Shotton, ed.), pages 55–68. Oxford: Clarendon Press.

Cooperative Holocene Mapping Project (COHMAP) members, 1988. Climatic changes of the last 18,000 years: Observations and model simulations. *Science* 241:1043–1052.

Cowling, S. A., 1999. Plants and temperature CO_2 uncoupling. *Science* 285:1500–1501.

Cox, P. M., R. A. Betts, C. D. Jones, S. A. Spall, and I. J. Totterdell, 2000. Acceleration of global warming due to carbon-cycle feedbacks in a coupled climate model. *Nature* 408:184–187.

Davis, M. B., 1986. Climatic instability, time lags, and community disequilibrium. In *Community Ecology* (J. Diamond and T. J. Case, eds.), pages 269–284. New York: Harper & Row.

Dawson, T. E., 1998. Fog in the California redwood forest: Ecosystem inputs and use by plants. *Oecologia* 117:476–485.

Dixon, R. K., S. Brown, R. A. Houghton, A. M. Solomon, M. C. Trexler, and J. Wisniewski, 1994. Carbon pools and flux of global forest ecosystems. *Science* 263:185–190.

Ehleringer, J. R., R. F. Sage, L. B. Flanagan, and R. W. Pearcy, 1991. Climate change and the evolution of C_4 photosynthesis. *TREE* 6(3):95–99.

Friedlingstein, P., P. Cox, R. Betts, L. Bopp, W. von Bloh, V. Brovkin, P. Cadule, S. Doney, M. Eby, I. Fung, G. Bala, J. John, C. Jones, F. Joos, T. Kato, M. Kawamiya, W. Knorr, K. Lindsay, H. D. Matthews, T. Raddatz, P. Rayner, C. Reick, E. Roeckner, K.-G. Schnitzler, R. Schnur, K. Strassmann, A. J. Weaver, C. Yoshikawa, and N. Zeng, 2006. Climate-carbon cycle feedback analysis: Results from the C^4MIP model intercomparison. *Journal of Climate* 19:3337–3353.

Grace, J., J. Lloyd, J. McIntyre, A. C. Miranda, P. Meir, H. S. Miranda, C. Nobre, J. Moncrieff, J. Massheder, Y. Malhi, I. Wright, and J. Gash, 1995. Carbon dioxide uptake by an undisturbed tropical rain forest in Southwest Amazonia, 1992 to 1993. *Science* 270:778–780.

Graumlich, L. J., 1991. Subalpine tree growth, climate, and increasing CO_2: An assessment of recent growth trends. *Ecology* 72(1):1–11.

Gribbin, J., 1990. *Hothouse Earth: The Greenhouse Effect and Gaia*. New York: Grove Weidenfeld.

Gruber, N., P. Friedlingstein, C. B. Field, R. Valentini, M. Heimann, J. E. Richey, P. Romero Lankao, E. D. Schultze, and C.-T. A. Chen, 2004. The vulnerability of the carbon cycle in the 21st century: An assessment of carbon-climate-human interactions. In *The Global Carbon Cycle: Integrating Humans, Climate, and the Natural World* (C. B. Field and M. R. Raupach, eds.), pages 45–76. Washington, D.C.: Island Press. A project of SCOPE (Scientific Committee on Problems of the Environment) of the International Council for Science.

Holdridge, L. R., 1947. Determination of world plant formations from simple climatic data. *Science* 105:367–368.

Houghton, R. A., J. L. Hackler, and J. T. Lawrence, 1999. The U.S. carbon budget: Contributions from land-use change. *Science* 285:574–578.

Jackson, R. B., J. L. Banner, E. G. Jobbágy, W. T. Pockman, and D. H. Wall, 2002. Ecosystem carbon loss with woody plant invasion of grasslands. *Nature* 418:623–626.

Jarvis, P., and S. Linder, 2000. Constraints to growth of boreal forests. *Nature* (Brief Communications) 405:904–905.

Keeling, H. C., and O. L. Phillips, 2007. The global relationship between forest productivity and biomass. *Global Ecology and Biogeography* 16:618–631.

Kimball, B. A., K. Kobayashi, and M. Bindi, 2002. Responses of agricultural crops to free-air CO_2 enrichment. *Advances in Agronomy* 77:293–368.

Knapp, A. K., J. M. Briggs, S. L. Collins, S. R. Archer, M. S. Bret-Harte, B. E. Ewers, D. P. Peters, D. R. Young, G. R. Shaver, E. Pendell, and M. B. Cleary, 2008. Shrub encroachment in North American grasslands: Shifts in growth form dominance rapidly alter control of ecosystem carbon inputs. *Global Change Biology* 14:615–623.

Kukla, G. J., M. L. Bender, J. L. de Beaulieu, G. Bond, W. S. Broecker, P. Cleveringa, J. E. Gavin, T. D. Herbert, J. Imbrie, J. Jouzel, L. D. Keigwin, K. L. Knudsen, J. F. McManus, J. Merkt, D. R. Muhs, and H. Müller, 2002. Last interglacial climates. *Quaternary Research* 58:2–13.

LaMarche, V. C., Jr., D. A. Graybill, H. C. Fritts, and M. R. Rose, 1984. Increasing atmospheric carbon dioxide: Tree ring evidence for growth enhancement in natural vegetation. *Science* 225:1019–1021.

Lovelock, J. E., 1979. *Gaia: A New Look at Life on Earth*, page 11. Oxford: Oxford University Press.

Lovelock, J. E., 2006. *The Revenge of Gaia: Why the Earth Is Fighting Back—And How We Can Still Save Humanity*. London: Allen Lane (imprint of Penguin Books).

Magnani, F., M. Mencuccini, M. Borghetti, F. Berninger, S. Delzon, A. Grelle, P. Hari, P. G. Jarvis, P. Kolari, A. S. Kowalski, H. Lankreijer, B. E. Law, A. Lindroth, D. Loustau, G. Manca, J. B. Moncrieff, V. Tedeschi, R. Valentini, and J. Grace, 2007. The human footprint in the carbon cycle of temperate and boreal forests. *Nature* 447:848–850.

McKibben, B., 2007. *Fight Global Warming Now: The Handbook for Taking Action in Your Community*. New York: Henry Holt. For more on the 350 campaign, see http://www.350.org.

Myneni, R. B., C. D. Keeling, C. J. Tucker, G. Asrar, and R. R. Nemani, 1997. Increased plant growth in the northern high latitudes from 1981 to 1991. *Nature* 386:698–702.

Norby, R. J., S. D. Wullschleger, C. A. Gunderson, D. W. Johnson, and R. Ceulemans, 1999. Tree responses to rising CO_2 in field experiments: Implications for the future forest. *Plant, Cell and Environment* 22:683–714.

Norse, E. A., 1990. *Ancient Forests of the Pacific Northwest*. Washington, D.C.: The Wilderness Society and Island Press.

Pagani, M., K. H. Freeman, and M. A. Arthur, 1999. Late Miocene atmospheric CO_2 concentrations and the expansion of C_4 grasses. *Science* 285:876–879.

Parrish, J. T., and R. A. Spicer, 1988. Middle Cretaceous wood from the Nanushuk Group, central North Slope, Alaska. *Paleontology* 31:19–34.

Parrish, J. T., I. L. Daniel, E. M. Kennedy, and R. A. Spicer, 1998. Paleoclimatic significance of mid-Cretaceous floras from the Middle Clarence Valley, New Zealand. *Palaios* 13:149–159.

Pearson, P. N., and M. R. Palmer, 1999. Middle Eocene seawater pH and atmospheric carbon dioxide concentrations. *Science* 284:1824–1826.

Pearson, P. N., and M. R. Palmer, 2000. Atmospheric carbon dioxide concentrations over the past 60 million years. *Nature* 406:695–699.

Pearson, P. N., P. W. Ditchfield, J. Singano, K. G. Harcourt-Brown, C. J. Nicholas, R. K. Olsson, N. J. Shackleton, and M. A. Hall, 2001. Warm tropical sea surface temperatures in the late Cretaceous and Eocene epochs. *Nature* 413:481–487.

Peters, R. L., III, 1988. The effect of global climate change on natural communities. In *Biodiversity* (E. O. Wilson and F. M. Peters, eds.), pages 450–461. Washington, D.C.: National Academy Press.

Petit, J.-R., J. Jouzel, D. Raynaud, N. I. Barkov, J.-M. Barnola, I. Basile, M. Bender, J. Chappellaz, M. Davis, G. Delaygue, M. Delmotte, V. M. Kotlyakov, M. Legrand, V. Y. Lipenkov, C. Lorius, L. Pepin, C. Ritz, E. Saltzman, and M. Stievenard, 1999. Climate and atmospheric history of the past 420,000 years from the Vostok ice core, Antarctica. *Nature* 399:429–436.

Phillips, O. L., Y. Malhi, N. Higuchi, W. F. Laurance, P. V. Nunez, R. M. Vasquez, S. G. Laurance, L. V. Ferreira, M. Stern, S. Brown, and J. Grace, 1998. Changes in the carbon balance of tropical forests: Evidence from long-term plots. *Science* 282:439–442.

Root, T. L., D. P. MacMynowski, M. D. Mastrandrea, and S. H. Schneider, 2005.

Human modified temperatures induce species change: Joint attribution. *Proceedings of the National Academy of Sciences* 102:7465–7469.

Ruddiman, W. F., 2001. *Earth's Climate: Past and Future*. New York: W. H. Freeman.

Shaw, M. R., T. E. Huxman, and C. P. Lund, 2005. Modern and future semiarid and arid ecosystems. In *A History of Atmospheric CO_2 and Its Effects on Plants, Animals and Ecosystems* (J. R. Ehleringer, M. D. Dearing, and T. E. Cerling, eds.), pages 415–440. Berlin: Springer.

Skelton, P. W., R. A. Spicer, S. P. Kelley, and I. Gilmour, 2003. *The Cretaceous World* (P. W. Skelton, ed.). Cambridge, U.K.: The Open University and Cambridge University Press.

Stephens, B. B., K. R. Gurney, P. P. Tans, C. Sweeney, W. Peters, L. Bruhwiler, P. Ciais, M. Ramonet, P. Bousquet, T. Nakazawa, S. Aoki, T. Machia, G. Inoue, N. Vinnichenko, J. Lloyd, A. Jordan, M. Heimann, O. Shibistova, R. L. Langenfelds, L. P. Steele, R. J. Francey, and A. S. Denning, 2007. Weak northern and strong tropical land carbon uptake from vertical profiles of atmospheric CO_2. *Science* 316:1732–1735.

Thomas, M. F., and M. B. Thorp, 1996. The response of geomorphic systems to climatic and hydrologic change during the late Glacial and early Holocene in the humid and sub-humid tropics. In *Global Continental Changes: The Context of Palaeohydrology* (J. Branson, A. G. Brown, and K. J. Gregory, eds.), pages 139–153. London: Geological Society Special Publication No. 115.

United Nations Food and Agricultural Organization, 2003. *State of the World's Forests*. Rome: U.N. Food and Agricultural Organization. Online at http://www.fao.org.

Webb, T., K. H. Anderson, P. J. Bartlein, and R. S. Webb, 1998. Late Quaternary climate change in eastern North America: A comparison of pollen-derived estimates with climate model results. *Quaternary Science Reviews* 17:587–606.

Weisman, A., 2007. *The World without Us*. New York: Thomas Dunne Books (imprint of St. Martin's Press).

Whitlock, C., and P. J. Bartlein, 1997. Vegetation and climate change in northwestern America during the past 125 kyr. *Nature* 388:57–61.

Wilson, R. C. L., S. A. Drury, and J. L. Chapman, 2000. *The Great Ice Age: Climate Change and Life*. London and New York: Routledge.

Wing, S. L., L. J. Hickey, and C. C. Swisher, 1993. Implications of an exceptional fossil flora for late Cretaceous vegetation. *Nature* 363:342–344.

Wolfe, J., 1993. A method of obtaining climate parameters from leaf assemblages. *U.S. Geological Survey Bulletin* 2040.

Zachos, J., M. Pagani, L. Sloan, E. Thomas, and K. Billups, 2001. Trends, rhythms, and aberrations in global climate 65 Ma to present. *Science* 292:686–693.

Zelikson, E. M., O. K. Borisova, C. V. Kremenetsky, and A. A. Velichko, 1998. Phytomass and carbon storage during the Eemian optimum, late Weichselian maximum and Holocene optimum in eastern Europe. *Global and Planetary Climate Change* 16–17:181–195.

Chapter 6. An Internal Cleanse

Belyea, L. R., and N. Malmer, 2004. Carbon sequestration in peatland: Patterns and mechanisms of response to climate change. *Global Change Biology* 10:1043–1052.

Bormann, B. T., H. Spaltenstein, M. H. McClellan, F. C. Ugolini, K. Cromack Jr., and S. M. Nay, 1995. Rapid soil development after windthrow disturbance in pristine forests. *Journal of Ecology* 83:747–757.

Brown, S., and A. E. Lugo, 1992. Aboveground biomass estimates for tropical moist forests of the Brazilian Amazon. *Interciencia* 17(1):8–18.

Davidson, E. A., and P. Artaxo, 2004. Globally significant changes in biological processes of the Amazon basin: Results of the Large-Scale Biosphere-Atmosphere Experiment. *Global Change Biology* 10:519–529.

Frolking, S., and N. T. Roulet, 2007. Holocene radiative forcing impact of northern peatland carbon accumulation and methane emissions. *Global Change Biology* 13(5):1079–1088.

Giardina, C. P., and M. G. Ryan, 2000. Evidence that decomposition rates of organic carbon in mineral soil do not vary with temperature. *Nature* 404:858–861.

Goulden, M. L., S. C. Wofsy, J. W. Harden, S. E. Trumbore, P. M. Crill, S. T. Gower, T. Fries, B. C. Daube, S.-M. Fan, D. J. Sutton, A. Bazzaz, and J. W. Munger, 1998. Sensitivity of boreal forest carbon balance to soil thaw. *Science* 279:214–216.

Greb, S. F., W. A. DiMichele, and R. A. Gastaldo, 2006. Evolution and importance of wetlands in earth history. In *Wetlands through Time* (S. F. Greb and W. A. DiMichele, eds.), pages 1–40. Boulder, CO: Geological Society of America Special Paper 399. (Citing J. C. McElwain, D. J. Beerling, and F. I. Woodward, 1999. Fossil plants and global warming at the Triassic-Jurassic boundary. *Science* 285:1386–1390.)

Gruber, N., P. Friedlingstein, C. B. Field, R. Valentini, M. Heimann, J. E. Richey, P. Romero Lankao, E. D. Schultze, and C.-T. A. Chen, 2004. The vulnerability of the carbon cycle in the 21st century: An assessment of carbon-climate-human interactions. In *The Global Carbon Cycle: Integrating Humans, Climate, and the Natural World* (C. B. Field and M. R. Raupach, eds.), pages 45–76. Washington, D.C.: Island Press. A project of SCOPE (Scientific Committee on Problems of the Environment) of the International Council for Science.

Gunnarsson, U., 2005. Global patterns of *Sphagnum* productivity. *Journal of Bryology* 27:269–279.

Jarvis, P., and S. Linder, 2000. Constraints to growth of boreal forests. *Nature* (Brief Communications) 405:904–905.

Jobbágy, E. G., and R. B. Jackson, 2000. The vertical distribution of soil organic carbon and its relation to climate and vegetation. *Ecological Applications* 10(2):423–436.

Keller, M., M. E. Mitre, and R. F. Stallard, 1990. Consumption of atmospheric methane in soils of central Panama: Effects of agricultural development. *Global Biogeochemical Cycles* 4(1):21–27.

Keller, M., R. Varner, J. D. Dias, H. Silva, P. Crill, R. Cosme de Oliveira Jr., and G. P. Asner, 2005. Soil-atmosphere exchange of nitrous oxide, nitric oxide, methane, and carbon dioxide in logged and undisturbed forest in Tapajos National Forest, Brazil. *Earth Interactions* 9(23):1–28.

Keppler, F., J. T. G. Hamilton, M. Braß, and T. Röckmann, 2006. Methane emissions from terrestrial plants under aerobic conditions. *Nature* 439:187–191.

Khalil, M. A. K., M. J. Shearer, and R. A. Rasmussen, 2000. Methane sinks, distribution and trends. In *Atmospheric Methane: Its Role in the Global Environment* (M. A. K. Khalil, ed.), pages 86–97. Berlin: Springer-Verlag.

Knapp, A. K., J. M. Briggs, S. L. Collins, S. R. Archer, M. S. Bret-Harte, B. E. Ewers, D. P. Peters, D. R. Young, G. R. Shaver, E. Pendell, and M. B. Cleary, 2008. Shrub encroachment in North American grasslands: Shifts in growth form dominance rapidly alter control of ecosystem carbon inputs. *Global Change Biology* 14:615–623.

Lugo, A. E., M. Brinson, and S. Brown (eds.), 1990. *Ecosystems of the World: Forested Wetlands.* Amsterdam: Elsevier.

Lugo, A. E., M. J. Sanchez, and S. Brown, 1986. Land use and organic carbon content of some subtropical soils. *Plant and Soil* 96:185–196.

Mitsch, W. J., and J. G. Gosselink, 2007. *Wetlands* (4th ed.). Hoboken, NJ: John Wiley & Sons.

Noe, G. B., and C. R. Hupp, 2005. Carbon, nitrogen, and phosphorus accumulation in floodplains of Atlantic coastal plain rivers, USA. *Ecological Applications* 15(4):1178–1190.

Parrish, J. T., A. M. Ziegler, and C. R. Scotese, 1982. Rainfall patterns and the distribution of coals and evaporates in the Mesozoic and Cenozoic. *Palaeogeography, Palaeoclimatology, Palaeoecology* 40:67–101.

Retallack, G. J., and E. S. Krull, 2006. Carbon isotopic evidence for terminal-Permian methane outbursts and their role in extinctions of animals, plants, coral reefs, and peat swamps. In *Wetlands through Time* (S. F. Greb and W. A. DiMichele, eds.), pages 249–268. Boulder, CO: Geological Society of America Special Paper 399.

Schlesinger, W. H., J. Palmer Winkler, and J. P. Megonigal, 2000. Soils and the global carbon cycle. In *The Carbon Cycle* (T. M. L. Wigley and D. S. Schimel, eds.), pages 93–101. Cambridge, U.K.: Cambridge University Press.

Skelton, P. W., R. A. Spicer, S. P. Kelley, and I. Gilmour, 2003. *The Cretaceous World* (P. W. Skelton, ed.). Cambridge, U.K.: The Open University and Cambridge University Press.

Sombroek, W., F. O. Nachtergaele, and A. Hebel, 1993. Amounts, dynamics and sequestering of carbon in tropical and subtropical soils. *Ambio* 22(7):417–426.

Tidwell, M., 2003. *Bayou Farewell: The Rich Life and Tragic Death of Louisiana's Cajun Coast*. New York: Vintage Departures (division of Random House).

Ward, P. D., 2006. *Out of Thin Air: Dinosaurs, Birds and Earth's Ancient Atmosphere*. Washington, D.C.: Joseph Henry Press.

Whiting, G. J., and J. P. Chanton, 2001. Greenhouse carbon balance of wetlands: Methane emissions versus carbon sequestration. *Tellus* 53B:521–528.

Chapter 7. Beneath the Surface

Berger, W. H., 1982. Increase of carbon dioxide in the atmosphere during deglaciation: The coral reef hypothesis. *Naturwissenschaften* 69:87–88.

Berner, R. A., 2004. *The Phanerozoic Carbon Cycle: CO_2 and O_2*. Oxford: Oxford University Press.

Berner, R. A., A. C. Lasaga, and R. M. Garrels, 1983. The carbonate-silicate geochemical cycle and its effect on atmospheric carbon dioxide over the past 100 million years. *American Journal of Science* 283:641–683.

Chan, F., J. A. Barth, J. Lubchenco, A. Kirincich, H. Weeks, W. T. Peterson, and B. A. Menge, 2008. Emergence of anoxia in the California Current large marine ecosystem. *Science* 319:920.

Dessert, C., B. Dupré, L. M. François, J. Schott, J. Gaillardet, G. Chakrapani, and S. Bajpai, 2001. Erosion of Deccan Traps determined by river geochemistry: Impact on the global climate and the $^{87}Sr/^{86}Sr$ ratio of seawater. *Earth and Planetary Science Letters* 188:459–474.

Gattuso, J.-P., M. Frankignoulle, and S. V. Smith, 1999. Measurement of community metabolism and significance in the coral reef CO_2 source-sink debate. *Proceedings of the National Academy of Sciences* 96(23):13017–13022.

Moulton, K. S., and R. A. Berner, 1998. Quantification of the effect of plants on weathering: Studies in Iceland. *Geology* 26:895–898.

Parrish, J. T., 1998. *Interpreting Pre-Quaternary Climate from the Geologic Record*. New York: Columbia University Press.

Raymo, M. E., 1994. The Himalayas, organic carbon burial, and climate in the Miocene. *Paleoceanography* 9:399–404.

Raymo, M. E., W. F. Ruddiman, and P. N. Froelich, 1988. Influence of late Cenozoic mountain building on ocean geochemical cycles. *Geology* 16:649–653.

Retallack, G. J., 2001. A 300-million-year record of atmospheric carbon dioxide from fossil plant cuticles. *Nature* 411:287–290.

Royer, D. L., 2006. CO_2-forced climate thresholds during the Phanerozoic. *Geochimica et Cosmochimica Acta* 70:5665–5675.

Schwartzmann, D. W., and T. Volk, 1989. Biotic enhancement of weathering and the habitability of Earth. *Nature* 340:457–460.

Skelton, P. W., R. A. Spicer, S. P. Kelley, and I. Gilmour, 2003. *The Cretaceous World* (P. W. Skelton, ed.). Cambridge, U.K.: The Open University and Cambridge University Press.

Chapter 8. Systematic Healing

Bala, G., K. Caldeira, M. Wickett, T. J. Phillips, D. B. Lobell, C. Delire, and A. Mirin, 2007. Combined climate and carbon-cycle effects of large-scale deforestation. *Proceedings of the National Academy of Sciences* 104(16):6550–6555.

Brazel, A., N. Selover, R. Vose, and G. Heisler, 2000. The tale of two climates—Baltimore and Phoenix urban LTER sites, *Climate Research* 15:123–135 (July 20).

Breshears, D. D., J. W. Nyhan, C. E. Heil, and B. P. Wilcox, 1998. Effects of woody plants on microclimate in a semiarid woodland: Soil temperature and evaporation in canopy and intercanopy patches. *International Journal of Plant Sciences* 159(6):1010–1017.

Gomez, F., J. Jabaloyes, and E. Vaño, 2004. Green zones in the future of urban planning. *Journal of Urban Planning and Development* June:94–100.

Harlan, S. L., A. J. Brazel, L. Prashad, W. L. Stefanov, and L. Larsen, 2006. Neighborhood microclimates and vulnerability to heat stress. *Social Science and Medicine* 63:2847–2863.

Hill, J. B., 2000. *The Legacy of Luna: The Story of a Tree, a Woman, and the Struggle to Save the Redwoods*, pages 76–77. San Francisco: HarperSanFrancisco (division of HarperCollins).

Jenerette, G. D., S. L. Harlan, A. Brazel, N. Jones, L. Larsen, and W. L. Stefanov, 2007. Regional relationships between surface temperature, vegetation, and human settlement in a rapidly urbanizing ecosystem. *Landscape Ecology* 22:353–364.

Kuo, F. E., and W. C. Sullivan, 2001. Environment and crime in the inner city: Does vegetation reduce crime? *Environment and Behavior* 33(3):343–367.

Lancaster, B., 2006. *Rainwater Harvesting for Drylands and Beyond.* Vol. 1: *Guiding Principles to Welcome Rain into Your Life and Landscape.* White River Junction, VT: Rainsource Press, distributed by Chelsea Green Publishing Company.

Lancaster, B., 2008. *Rainwater Harvesting for Drylands and Beyond.* Vol. 2: *Water-Harvesting Earthworks.* White River Junction, VT: Rainsource Press, distributed by Chelsea Green Publishing Company.

Makarieva, A. M., V. G. Gorshkov, and B. L. Li, 2006. Conservation of water cycle on land via restoration of natural closed-canopy forest: Implications for regional landscape planning. *Ecological Research* 21:897–906.

Mitsch, W. J., and J. W. Day Jr., 2006. Restoration of wetlands in the Mississippi-Ohio-Missouri (MOM) River Basin: Experience and needed research. *Ecological Engineering* 26:55–69.

Mueller, E. C., and T. A. Day, 2005. The effect of urban ground cover on microclimate, growth and leaf gas exchange of oleander in Phoenix, Arizona. *International Journal of Biometeorology* 49:244–255.

Orr, D. W., 2006. *Design on the Edge: The Making of a High-Performance Building.* Cambridge, MA: MIT Press.

Riley, J. J., N. G. Hicks, and T. L. Thompson, 1992. Effect of Kuwait oil field fires on human comfort and environment in Jubail, Saudi Arabia. *International Journal of Biometeorology* 36:36–38.

Tiner, R. W., 1998. *In Search of Swampland: A Wetland Sourcebook and Field Guide*. New Brunswick, NJ: Rutgers University Press.

Chapter 9. Conclusion

Gore, Al, 2006. *An Inconvenient Truth: The Planetary Emergency of Global Warming and What We Can Do about It*. New York: Rodale and Melcher Media.

Guggenheim, Davis (director), 2006. *An Inconvenient Truth: A Global Warning*. Global Warming Documentary, LLC, and Paramount Pictures.

Law, B. E., D. Turner, J. Campbell, O. J. Sun, S. Van Tuyl, W. D. Ritts, and W. B. Cohen, 2004. Disturbance and climate effects on carbon stocks and fluxes across western Oregon, USA. *Global Change Biology* 10:1429–1444.

Lenart, M., 2006. Collaborative stewardship to prevent wildfires. *Environment* 48(7):8–21.

Strom, B. A., and P. Z. Fule, 2007. Pre-wildfire fuel treatments affect long-term ponderosa pine forest dynamics. *International Journal of Wildland Fire* 16:128–138.

Stuever, M., 2009. *The Forester's Log: Musings from the Woods*. Albuquerque: University of New Mexico Press.

Verchick, R. R. M., 2010. *Facing Catastrophe: Environmental Action for a Post-Katrina World*. Cambridge, MA: Harvard University Press.

Wiedinmyer, C., and J. C. Neff, 2007. Estimates of CO_2 from fires in the United States: Implications for carbon management. *Carbon Balance and Management*, published online November 1, 2007: http://www.cbmjournal.com/content/2/1/10<\>

Acknowledgments

So many people helped bring this book to completion that it's difficult to name them all. The loving support of my spouse, Bob Segal, kept me going through many challenges. And his assistance on some of the many tasks involved, such as entering the edits I made on hard copy in coffee shops, helped lighten my load tremendously. He was also one of many readers who helped troubleshoot the text. Others included Lynn Jacobs, Anna Mirocha, Eric van Lienden, Kay Sather, Sonya Diehn, Richard Druitt, Teala Smith, Dawn Madden, Susan Lenart-Kazmer, Jeneiene Schaffer, and Lucia Grossberger Morales. Lucia Grossberger Morales also graciously provided her solar-powered home in Tucson for my use as a writing getaway while she traveled. Richard Druitt and Andria Robinson similarly lent me their lovely riverside home in Puerto Rico for more than a month while I fed the cat, watered plants, and wrote. Occasional immersions in the Caribbean Sea were exactly what I needed to move forward on this project, which initially felt overwhelming to someone more accustomed to writing news stories than the seemingly never-ending string of pages that comprise a book.

Getting a contract to write the book, though, was something I had dreamed about for years. I thank Gary Heidt, co-founder of Signature Literary Agency, for helping me reach that dream. Allyson Carter, too, has been an essential guiding force in her role as editor for the University of Arizona Press. I also thank Debra Makay for her professional editing that greatly improved the manuscript.

The idea of writing this book has been on my mind for a long time. It's what drew me back to graduate school in the mid-1990s, when I realized I would not reach the depth I wanted without deep background. So finishing it feels like the culmination of that lengthy Ph.D. process more so than finishing my dissertation did. I thank the many mentors who guided me in my graduate work, including Sandra Brown, Ariel Lugo,

Fred Scatena, Waite Osterkamp, Steve Leavitt, Judy Parrish, Malcolm Hughes, and Gregg Garfin. All played important roles in my graduate (and in Gregg Garfin's case, postdoctoral) education. I also thank all the scientists and other knowledgeable sources who generously spent time talking with me and double-checking my facts. For lack of space, I ended up leaving out too many juicy bits in my final edit, but all the conversations shaped the book's direction. I also appreciated the insights and support of two anonymous reviewers.

Many instructors at Tucson's Yoga Connection, especially Priscilla Potter, Ashley Leal, and Tracy Gordon, helped me find the physical stamina to sit at the computer for so many long hours and the mental stillness to focus my thoughts. Finally, I am grateful that my dog, Pepita, pulled through a near-fatal illness so she could sit at my feet throughout the writing of this book. There's nothing like the love of a member of another species to remind us that this planet holds mysteries beyond our human experience.

Index

About the Author

Melanie Lenart, Ph.D., is an environmental scientist and writer who specializes in climate change and forests. As a scientist, she studied forest dynamics in China, Colorado, and Puerto Rico, where she lived during two major hurricanes. She was involved in an Arizona agricultural experiment testing how plants responded to elevated levels of carbon dioxide, the main greenhouse gas responsible for the ongoing warming of the planet. Her research with the world-renowned University of Arizona Laboratory of Tree-Ring Research included dating logs to compare their decay state and soil dynamics. While working as a postdoctoral researcher with the University of Arizona's Climate Assessment for the Southwest (CLIMAS), she researched forest policy in the aftermath of an Arizona wildfire that torched nearly half a million acres. Some of the many feature articles she wrote for CLIMAS have been pulled into a book compilation, *Global Warming in the Southwest.* An award-winning journalist, Lenart worked as an environmental writer for Puerto Rico's daily *San Juan Star* in the mid-1990s. She lives with her husband in Tucson, where she teaches environmental writing and writes about the many facets of climate change and its impacts—including what we can do about it.